CHAOS IN NONLINEAR DYNAMICAL SYSTEMS

Proceedings of a Workshop held at
The U.S. Army Research Office
Research Triangle Park, North Carolina
March 13–15, 1984

The workshop was sponsored by the
Mathematical Sciences Division
United States Army Research Office
Research Triangle Park, North Carolina

Chaos in Nonlinear Dynamical Systems

edited by
Jagdish Chandra
U.S. Army Research Office

siam Philadelphia/1984

Library of Congress Catalog Card Number: 84-52603
ISBN: 0-89871-052-9

CONTENTS

INTRODUCTION

There is a resurgence of research activity in nonlinear dynamics. Significant progress in qualitative techniques for finite and infinite dimensional systems together with the astronomical advances in computational powers is enabling better understanding of complex phenomena. Such progress until now was either unattainable or too cumbersome to attempt. In recent years, specifically, a great deal of progress has been made in the study of stability and chaotic behavior of nonlinear systems. Still many more challenging problems await resolution. The keen interest in this field is primarily due to the fact that the satisfactory solution of such problems is bound to have significant impact on a wide variety of physical and engineering applications. These include multi-stable chemical and biological systems, electronic and optical switches, structure of turbulence, oscillation in aeroelastic structures, feedback controls and communication and power systems.

Recognizing the importance of this field, the Mathematical Sciences Division, U.S. Army Research Office organized an interdisciplinary workshop on Chaos in Nonlinear Dynamical Systems. The workshop was held on March 13–15, 1984 at the Army Research Office, Research Triangle Park, North Carolina, attended by more than sixty mathematicians, engineers, physicists and chemists. This book is based on the papers presented at this workshop.

In Chapter I, Newhouse discusses various mathematical notions useful in the description of chaotic motions. Pseudorandom phenomena have recently been shown to be relevant to a variety of engineering applications in place of stochastic models commonly used in communication and control theory, and structural analysis. The problem of generating pseudorandom processes using models which are sufficiently simple so as to allow some quantitative analysis is, therefore, drawing much attention. In Chapter II, Brockett and Cebuhar, describe a piecewise linear third order autonomous differential equation to seek new insights into particularly simple mechanisms which appear to generate chaos in such systems.

The next three chapters are devoted to the development and application of the Poincaré-Melnikov-Arnold method. For instance, in Chapter III, Marsden proves existence of chaos in the sense of Poincaré-Birkhoff-Smale horseshoes, followed in the next chapter by Slemrod, who discusses the application of this approach to chaos to the problem of equilibrium distribution of a van der Waals fluid undergoing spatially thermal variations. Using similar techniques, in Chapter V, Salam and Sastry describe the complete dynamics of the Josephson junction circuit and provide a complete bifurcation diagram of the a.c. forced junction. This is done by establishing analytically the existence of chaos for certain parameter ranges in the dynamics of such systems.

In Chapter VI, Levi considers the so-called beating modes in Josephson junction circuits. He provides qualitative analysis of these systems and gives topological characterization of these modes. The perturbed Sine-Gordon equation models the dynamics of long Josephson tunnel junction. In Chapter VII, Christiansen discusses some of the soliton dynamic states for this equation and describes computer experiments exhibiting hysteresis phenomena and chaotic intermittency between soliton dynamics states occurring as a result of applied external bias. In the following chapter, Kopell discusses the influence of symmetry properties on coherence and chaos in a chain of weakly coupled oscillations. This study is partly motivated by a biological application.

Existence of multiple stable states has importance in a variety of applications. Studies in optical bistability, for instance, have opened up new exciting possibilities of using such logic elements as switches in optical computing. In Chapter IX, McLaughlin, Moloney and

Newell summarize some recent results on coherence and chaos in optical bistable laser cavities. Next, Moss explores two very different types of switches and their behavior when they are subject to large amplitude external interference. In Chapter XI, Arecchi investigates multistability and chaos in quantum optics. Three experimental situations are described that exemplify onset of chaos. Next, Ackerhalt and Milonni describe how chaotic dynamics of molecular vibrations explain the dependence of multi-photon absorption on pulse energy rather than intensity. In Chapter XIII, Casati and Guarneri investigate the chaotic properties of quantum systems under external perturbations.

In the last two chapters, the authors discuss implications of chaotic dynamics to turbulent flows and oscillation in mechanical systems. In Chapter XIV, Manley shows how recent research on the asymptotic properties of the Navier-Stokes equation is valuable in the use of computers as experimental tools in the study of the dynamics of fluids. He shows how the conventional estimate of the number of degrees of freedom of turbulent flows can be obtained from such asymptotic properties. In the last chapter, Dowell and Pierre look at the fundamental mechanisms in nonlinear mechanics that lead to chaotic oscillations. Through specific examples, they identify two categories of systems. In the first category the chaotic oscillations arise as a result of instability of the system to large finite disturbance while in the second category the chaos results from instability with respect to infinitesimal disturbances.

I thank all of those whose efforts have helped both to make the workshop successful and to bring this book into its final form. I particularly appreciate the contributors to this volume. Special thanks are due to Mrs. Frances Atkins, Mrs. Mary Mitchell and Mrs. Brenda Hunt of the Army Research Office for the diligent cooperation through all phases of this workshop and the preparation of this book.

JAGDISH CHANDRA
September 1984

UNDERSTANDING CHAOTIC DYNAMICS

S. E. NEWHOUSE*

Abstract: Various mathematical notions useful in the description of chaotic motion are discussed. Emphasis is given to hyperbolic attractors and Bowen-Ruelle-Sinai measures.

We describe here certain mathematical structures, results, and methods which are useful for understanding chaotic dynamical systems. Roughly speaking, a chaotic dynamical system is one which has presumably many) solutions which display highly aperiodic or erratic time dependence. A glance at recent physical and engineering literature (e.g. Holmes[3], Swinney[8]) reveals an abundance of physical systems with such motions. For reasons of space we will mainly deal with discrete dynamical systems (i.e. mappings from a subset M of Euclidean space to itself). Most of the results we discuss here have counterparts for systems with continuous time. For more information on this subject as well as related ideas, we refer to Newhouse[4,5,6,7], Farmer et al[1], and Guckenheimer and Holmes[2].

We consider a smooth manifold M (possibly with boundary) and a twice differentiable mapping f from M to M. We assume f is one-to-one and the inverse map $f^{-1}: f(M) \to M$ is also twice

---

*Department of Mathematics, University of North Carolina, Chapel Hill, NC 27514

differentiable. Given $x \in M$, consider the positive orbit $O_+(x) = \{x, f(x), f^2(x), \ldots\}$ of x and the ω-limit set $\omega(x) = \omega(x,f) = \{y \in M$: there is a sequence $n_1 < n_2 < \ldots$ such that $f^{n_i}x \to y$ as $i \to \infty\}$. Note that if $O_+(x)$ is bounded, then $\omega(x)$ is a closed, bounded set, and $f(\omega(x)) = \omega(x)$. In general, we are interested in describing $\omega(x,f)$ for as many x and f as possible. A simple situation arises when each $\omega(x,f)$ is a periodic orbit for $x \in M$. Then each initial state tends toward a periodic orbit. Such an f is not chaotic.

A first instance in which f might be called chaotic is when there are uncountably many points x for which $\omega(x,f)$ is also uncountable. This can occur for relatively mild f: let S^1 be the unit circle in $\mathbb{R}^2$, and let $f = S^1 \to S^1$ be defined by $f(u,v) = (u \cos 2\pi\alpha - v \sin 2\pi\alpha,\ u \sin 2\pi\alpha + v \cos 2\pi\alpha)$ for $(u,v) \in S^1$ and α irrational. Thus, f is just a rotation through angle $2\pi\alpha$. It is well-known that $\omega((u,v),f) = S^1$ for each $(u,v) \in S^1$. An additional condition frequently encountered in chaotic systems is "sensitive dependence on initial conditions." This could be defined as follows. Let $d(x,y)$ denote the distance between x and y. A point x exhibits sensitive dependence on intitial conditions if there are an $\alpha > 0$ and a constant $C > 0$ such that for any $\varepsilon > 0$ and any positive integer $n > 0$, there is a point y such that

(1) $d(x,y) < \varepsilon$

(2) $d(f^j x, f^j y) \geq Ce^{\alpha j} d(x,y)$ for $0 \leq j \leq n$.

The infinitesimal version of this is more easily defined: there is a vector v tangent to M at x such that $\limsup_{n \to \infty} \frac{1}{n} \log |T_x f^n v| > 0$. Here $T_x f^n$ is the derivative (linear approximation) to f^n at x and $|w|$ the norm of a vector w. The latter condition on x and v is frequently referred to as "(x,v) has a positive Lyapunov exponent." Sometimes one says that x has a positive Lyapunov exponent if there is a v such that (x,v) has one. Again there are simple mappings which satisfy this condition. Let $f: \mathbb{R}^2 \to \mathbb{R}^2$ be defined by $f(u,v) = (\lambda^{-1}u, \lambda v)$ with $1 < \lambda$. Any point in $\mathbb{R}^2$ has a positive exponent. All the points in $\mathbb{R} \times \{0\}$ have in addition, bounded positive orbits.

If we put the above definitions together we get a better definition of a chaotic f: say $f: M \to M$ is <u>chaotic</u> if there is a closed bounded subset V of M with non-empty interior such that

(1) f maps V into its interior .

(2) there are uncountably many points x in V for which $\omega(x)$ is uncountable and x has a positive Lyapunov exponent.

In the known examples where f is chaotic in this last sense and the chaos is persistent in the sense that any g C^1 near f also satisifies (1) and (2), one has some interesting behavior for f. For instance, f must have infinitely many periodic orbits in V, and f must have transverse homoclinic points (see Newhouse[4)], and Guckenheimer-Holmes[2)]) in V.

We wish to consider further notions of chaos. There is the frequently used term "strange attractor." This term was created by

Ruelle and Takens[9], and has been modified by various authors subsequently. One could define such an object as follows. A closed set $\Lambda \subset M$ is <u>invariant</u> if $f(\Lambda) = \Lambda$. The set Λ is an <u>attractor</u> if there is an open set $U \supset \Lambda$ such that $f(U) \subset U$ and $\bigcap_{n\geq 0} f^n(U) = \Lambda$. The attractor is <u>strange</u> if it has a complicated topology; e.g. if it is not a manifold. Frequently, one adds further indecomposability conditions as well. One such condition is that there is a point x in Λ whose orbit is dense in Λ. The basin of the attractor Λ is the set $B(\Lambda) = \bigcap_{n\leq 0} f^n(U) = \{x \in M: \omega(x) \subset \Lambda\}$. The idea of Ruelle and Takens was that typical points in the basin of a strange attractor would have a complicated asymptotic behavior, and that such objects could provide models for chaotic motion. In particular, the continuous time versions of such objects in certain infinite dimensional spaces could provide models for turbulence in fluids. Experiments in a number of situations have lent credence to this idea. Our above notions of chaos are independent of the notion of a strange attractor but they are certainly consistent with it: one merely has to consider V as a subset of $B(\Lambda)$.

In addition to (1) and (2) above, there are further conditions one could put on the set V. For instance, one might require the points x in (2) to be dense in V or to have positive or full volume in V. If the latter condition holds, one could ask for a description of statistical properties of $0_+(x)$ for typical x in V. For example, how are the points $\{x, fx, \ldots, f^n x\}$ distributed for typical x in V and large n? What is the structure of time averages $\frac{1}{n} \sum_{k=0}^{n-1} \psi(f^k x)$ for continuous ψ, large n, and typical x? These questions and many more have been answered in the case when V is in the basin of attraction of a hyperbolic attractor. Such objects provide mathematically understood models of chaotic behavior. A large amount of current research in dynamics attempts to extend the mathematics of hyperbolic attractors to more general situations. See Newhouse[7)] for related ideas.

It is our contention that much can be gained toward understanding chaotic dynamics by the careful study of hyperbolic attractors. That is, many of the phenomena arising in such attractors also occur in many other chaotic situations. There are several rigorous results supporting this contention although much work remains to be done.

Let us now recall the notion of a hyperbolic attractor, and give some typical examples.

A closed bounded invariant set Λ is <u>hyperbolic</u> if there are constants $C > 0$, $0 < \lambda < 1$, and for each $x \in \Lambda$, there is a splitting $T_x M = E_x^s \oplus E_x^u$ such that

(a) $v \in E_x^s \Longrightarrow T_x f(v) \in E_{f(x)}^s$ and

$v \in E_x^u \Longrightarrow T_x f(v) \in E_{f(x)}^u$

(b) $n \geq 0$ and $v \in E_x^s \Longrightarrow |T_x f^n(v)| \leq C\lambda^n |v|$ and

$n \geq 0$ and $v \in E_x^u \Longrightarrow |T_x f^{-n}(v)| \leq C\lambda^n |v|$.

Condition (a) describes invariance of the infinitesimal subspaces E_x^s and E_x^u, and condition (b) describes exponential forward contraction of vectors in E_x^s and exponential backward contraction of vectors in E_x^u. An attractor which is also a hyperbolic set is a <u>hyperbolic attractor</u>. Let us give some examples.

1. Let $\mathbb{R}^2 = \{(u,v)\}$ be the Euclidean plane, and let

   $A(u,v) = (2u + v, u + v)$.

   Thus A is a linear map with matrix $\binom{21}{11}$, determinant 1, and two real eigenvalues $\lambda = \frac{3 + \sqrt{5}}{2}$, $\lambda^{-1} = \frac{3 - \sqrt{5}}{2}$, with eigenspaces E_λ, and $E_{\lambda^{-1}}$, respectively. Let $T^2 = \mathbb{R}^2/Z^2$ be the two-dimensional torus where Z^2 is the integer lattice in $\mathbb{R}^2$. The map A induces a diffeomorphism $\bar{A}$ from T^2 to itself. Let $\pi: = \mathbb{R}^2 \to T^2$ be the natural projection. Let $x \in T^2$ and let $\bar{x} \in \mathbb{R}^2$ be such that $\pi(\bar{x}) = x$. Let $E_\lambda(\bar{x})(E_{\lambda^{-1}}(\bar{x}))$ be the line in $\mathbb{R}^2$ through $\bar{x}$ parallel to $E_\lambda(E_{\lambda^{-1}})$. Let $E_x^s = \pi(E_{\lambda^{-1}}(\bar{x}))$ and $E_x^u = \pi(E_\lambda(\bar{x}))$. It is easily checked that E_x^s and E_x^u are independent of the choice of $\bar{x} \in \pi^{-1}(x)$ and satisfy conditions (a) and (b) with $\bar{A} = f$. Then, all of T^2 is a hyperbolic attractor, and $B(T^2) = T^2$.

2. Let $\mathbb{C}$ be the set of complex numbers. Let

   $D = \{w \in \mathbb{C} = |w| \leq 1\}$, and let $S^1 = \{z \in \mathbb{C}: |z| = 1\}$.

   Thus, S^1 is the unit circle and D is the closed unit disk. Let $S^1 \times D$ be the product space which we think of as a solid torus of revolution in $\mathbb{R}^3$. Letting (x_1,y_1,z_1) be coordinates in $\mathbb{R}^3$, we consider $S^1 \times D =$

   $$\{(x_1,y_1,z_1): (x_1^2 + y_1^2)^{1/2} - a)^2 + z_1^2 \leq b^2\}$$

where $0 < b < a$, and $(x_1, y_1, 0)$ corresponds to $(x_1 + \sqrt{-1}\, y_1, 0)$ in $S^1 \times D$. The mapping $f(z,w) = (z^2, \frac{z}{2} + \frac{w}{4})$ maps $S^1 \times D^2$ into its interior as in the next figure.

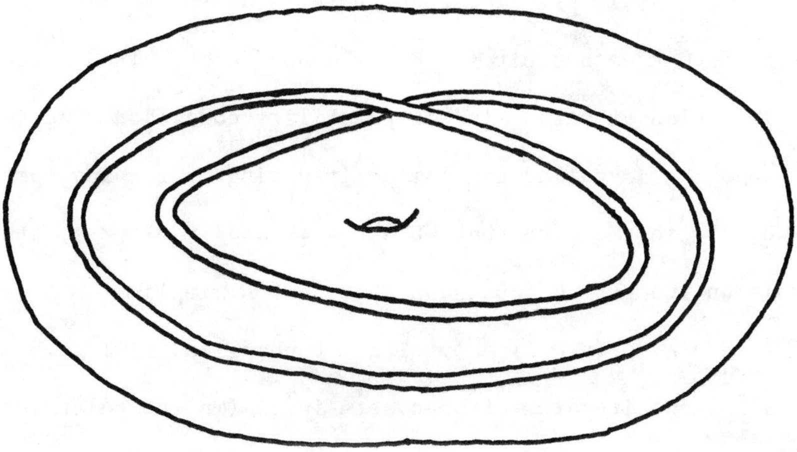

The largest invariant set $\Lambda = \bigcap_{n \geq 0} f^n(S^1 \times D)$ is a hyperbolic attractor of topological dimension 1. In fact, it looks locally like the product of a Cantor set and an interval.

3. Geodesic (inertial motion on a surface of negative curvature is such that the motion on each positive energy surface is the continuous time analog of a hyperbolic attractor.

Theorem 1 below presents some statistical results which have been obtained for hyperbolic attractors.

Recall that a Borel probability measure μ on M is

a non-negative countably additive real-valued set function defined on σ-field of Borel sets in M such that $\mu(M) = 1$. The measure μ is invariant if $\mu(f^{-1}A) = \mu(A)$ for every Borel set A. One calls μ ergodic if whenever E is a Borel set such that $f(E) = E$ it follows that $\mu(E) = 0$ or 1. If μ is ergodic, then there is a set E with $\mu(E) = 1$ such that each orbit in E is dense in E. Thus, ergodicity is a kind of strong indecomposability condition. We recall that Λ is called topologically mixing if for any open sets U, V in M such that $U \cap \Lambda \neq \phi$ and $V \cap \Lambda \neq \phi$, there is an integer $N > 0$ such that $n \geq N$ implies $f^n(U \cap \Lambda) \cap (V \cap \Lambda) \neq \phi$. Thus, topological mixing means that all large iterates of open sets in Λ (in the relative topology) get spread around well in Λ. A hyperbolic attractor containing a fixed point is, in fact, topologically mixing. We will say that the hyperbolic attractor is non-trivial if it contains more than one orbit. It follows that it must be uncountable and even have Hausdorff dimension greater than or equal to 1.

<u>Theorem 1</u> Assume $f\colon M \to M$ has a topologically mixing non-trivial hyperbolic attractor. Then,

(a) there is a set $A \subset B(\Lambda)$ such that the Lebesque measure of $B(\Lambda) - A$ is zero, and for every $x \in A$ and every continuous function $\phi\colon M \to \mathbb{R}$, $\lim_{n\to\infty} \frac{1}{n} \sum_{k=0}^{n-1} \phi(f^k x) = \hat{\phi}(x)$ exists and is independent of x in A

(b) the operator $\phi \longrightarrow \mu(\phi) = \hat{\phi}(x)$ for $x \in A$ determines an invariant measure called the Bowen-Ruelle-Sinai measure (or natural measure[1]) for Λ.

(c) μ is exponentially mixing: given C^1 real-valued functions ϕ and ψ, $\left|\int \phi(f^n x)\, \psi(x)\, d\mu - \int \phi d\mu \cdot \int \psi d\mu\right| \le Ce^{-n\alpha}$

for $n \ge 0$ and some constant $C > 0$.

(d) μ satisfies a central limit theorem[11]: for typical C^1 ψ there is a constant $\sigma = \sigma(\psi) > 0$ so that

$$\mu\{x \in V: \frac{1}{\sqrt{n}} (\sum_{k=0}^{n-1} \psi(f^k x) - n \int \psi du) < r\}$$

$$\longrightarrow \frac{1}{\sigma\sqrt{2\pi}} \int_{-\infty}^{r} \exp(-x^2/2\sigma^2)\, dx \quad \text{as} \quad n \to \infty.$$

(e) μ is "stable" under random perturbations[12].

Parts (a), (b), (c) of theorem 1 are due to Ruelle[10].

Remarks:

1. There are instances of chaotic f with no non-trivial hyperbolic attractors. However, the properties in (a) through (e) of theorem 1 may still hold in much generality. Even in the absence of rigorous proofs, theorem 1 suggests numerical tests which can be applied. For instance, one could compute time averages of various fuctions at various points to see if limiting time averages exist. Given a

function $\psi: M \to \mathbb{R}$, and an integer $n > 0$, one could compute

$$A(n,x,\psi) = \frac{1}{i} \sum_{k=0}^{i-1} \psi(f^{n+k}x)\, \psi(f^{k}x)) - (\frac{1}{i} \sum_{k=0}^{i-1} \psi(f^{k}x))^2$$

for typical x and i much larger than n. Suppose this were nearly independent of x and large i and decreased exponentially with n. Then, one would have evidence of the existence of a measure satisfying (a) and (c). Similarly, one could test for (d).

2. There are necessary and sufficient conditions for the existence of measures (with non-zero Lyapunov exponents) satisfying (a) for points x in sets of positive Lebesque measure. Precise statements are found in Newhouse[7] and the proofs will appear in the future.

References

1. D. Farmer, E. Ott, and J. Yorke, "The Dimension of Chaotic Attractors," Physica 7D (1983), 153.

2. J. Guckenheimer and P. Holmes, Non-linear Oscillations, Dynamical Systems, and Bilfurcations of Vector Fields, Applied Math. Sci. 42, Springer-Verlag, 1983.

3. P. Holmes, New Approaches to Non-linear Problems in Dynamics, SIAM, 1980.

4. S. Newhouse, Lectures on Dynamical Systems, Progress in Math 8, Birkhauser, Boston (1980), 1-115.

5. ___________, Chaotic Behavior of Deterministic Systems, ed. G. Iooss et al, Les Houches XXXVI (1981), 434-442.

6. ____________, Entropy and smooth dynamics, Dynamical Systems and Chaos, lecture Notes in Physics 179 (1983), Springer-Verlag, 165-180.

7. ___________, Probabilistic Ideas in Smooth Dynamical Systems, Proc. Conf. on Group Theoretical Methods in Physics, Univ. of Md., College Pk., Md., May, 1984.

8. H. Swinney, N. B. Abraham, and J. P. Gollub, Meeting Report, Testing Non-linear Dynamics, Physica 11D (1984), 252-264.

9. D. Ruelle and F. Takens, On the Nature of Turbulence, Comm. Math. Phys. 20 (1971), 167-192.

10. D. Ruelle, A Measure Associated with Axiom A Attractors, Amer. Jour. Math. 98 (1976), 619-654.

11. M. Ratner, "The central limit theorem for geodesic flows on n-dimensional manifolds of negative curvature," Isreal J. Math 16 (1973), 181-197.

12. Y. Kifer, "General random perturbations of hyperbolic and expanding transformations, pre-print, The Hebrew University of Jerusalem, June, 1983.

A PROTOTYPE CHAOTIC DIFFERENTIAL EQUATION

ROGER W. BROCKETT* AND WENCESLAO CEBUHAR**

Abstract. In this note we describe a piecewise linear third order autonomous differential equation which has, in a certain section of its state space, a first return map which is very well approximated by a unimodal map of an interval into itself. This map is qualitatively and quantitatively very much like that of the logistic equation. It is, in some respects, more convenient than the well-known example of Rössler [1] because it is piecewise linear. We go on in the last section to analyze the problem of making still simpler strange attractors.

1. Introduction. Engineers use stochastic process models in communication theory, control theory, structural analysis, analysis of vibrations, etc. In many cases the models used are gauss-markov even though there is little reason to believe such models are the appropriate ones. In fact, pseudorandom phenomena have recently been shown to be relevant to a variety of engineering situations, providing some support to the idea that we will now begin to see a shift away from the arbitrary use of stochastic modeling. In this note we address the problem of generating pseudorandom processes using models which are sufficiently simple so as to allow some quantitative analysis. While we do not solve the problem of proving that our systems have a strange attractor, we do develop some new insights into the question of suspending first return maps of the type which give rise to chaos.

*Division of Applied Sciences, Harvard University, Cambridge, MA 02138.
Work supported in part by the U.S. Army Research Office under Grant No. DAAG29-79-C-0147 and the National Science Foundation under Grant No. ECS-81-21428.
**Division of Applied Sciences, Harvard University, Cambridge, MA 02138. Work supported by Consejo Nacional de Investigaciones Cientificas y Técnicas, Argentina.

2. <u>The System</u>. The system under discussion here is

$$\dddot{y} + \ddot{y} + 1.25\,\dot{y} + f(y) = 0$$

with f being piecewise linear and of the form

$$f(y) = \begin{cases} \alpha y & ; \ |y| \leq 1 \\ \beta y - (\operatorname{sgn} y)(\beta - \alpha); & |y| \geq 1 \end{cases}$$

More specifically, we let $\alpha = -1.8$ and $\beta = 3.33$.

This is a system of the same general form as was described earlier by Sparrow [2] and by Brockett [3] and Brockett and Loncaric [4]. In [2] the role of homoclinic equilibrium points was emphasized. In this paper we investigate a parameter range -- one in which is not dominated by homoclinic phenomena. In fact, we have deliberately looked for a range of parameters for which the (apparent) strange attractor is very strongly localized and as simple as possible. Even though the parameter values are close to those explored in [3] and [4] we see quite different, and in some sense more interesting, phenomena.

We refer to [3] for a more detailed qualitative description of the trajectories associated with this differential equation. The main features of the solutions are explained by the fact that there is, for $|y| > 1$ two unstable complex eigenvalues and one stable real one whereas in the region $|y| \leq 1$ there is one unstable real eigenvalue and two stable complex eigenvalues. In the region $|y| > 1$ the modes associates to the complex eigenvalues tend to force the motion into a circle thus returning it to the region $|y| \leq 1$. On the other hand, in the region $|y| \leq 1$ the unstable real eigenvalue tends to produce a screw motion along an axis defined by the eigenvector associated with the unstable eigenvalue. This causes trajectories to leave $|y| \leq 1$ in the general direction of this eigenvector. The overall motion is then confined by the "caps" in the region $|y| > 1$ and is prevented from settling down in the region $|y| < 1$ by the unstable real eigenvalue. All this occurs for a range of parameters which is rather wide compared with the more specific phenomenon which we are about to describe.

Consider the trajectory shown in Figure 1. It is qualitatively consistent with the above discussion but has additional structure. It is, for example, not symmetric even though the equations of motion are invariant under the transformation $y \to -y$. Figures 2 and 3 show a scatter diagram consisting of the points on a long trajectory (approximately 10,000 units of time) as they pass through the plane $y=1$. Figure 2 corresponds to the passage with $\dot{y} > 0$ and Figure 3 corresponds to the passage with $\dot{y} < 0$. These are, then, the cross sections of the (apparent) strange attractor.

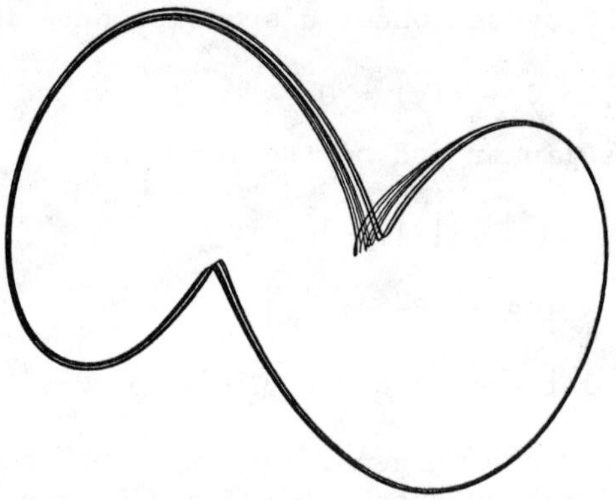

Figure 1. A sample trajectory

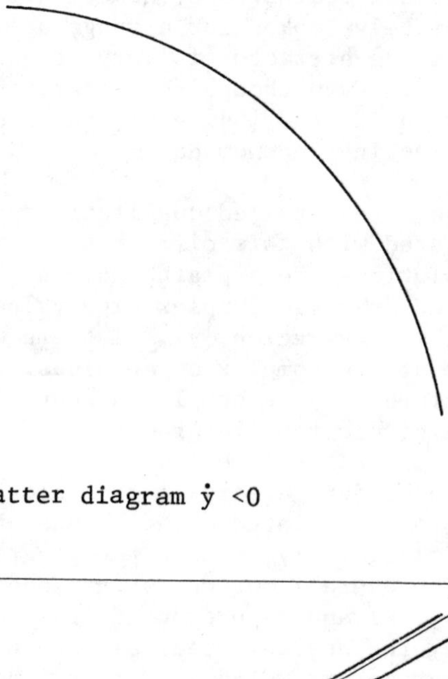

Figure 2. Scatter diagram $\dot{y} < 0$

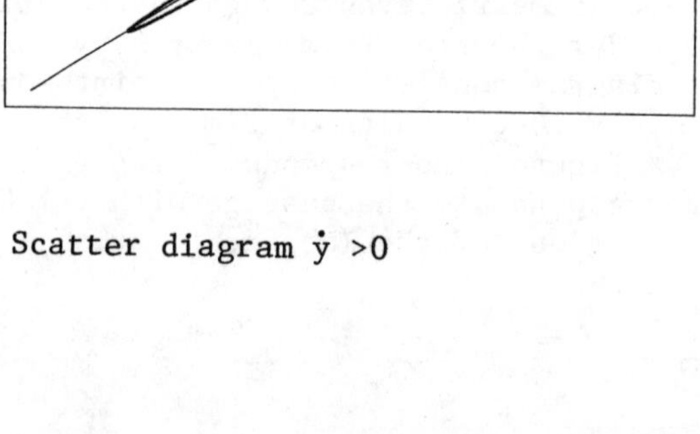

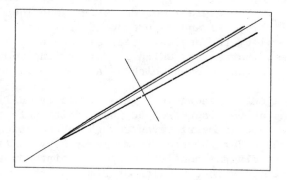

Figure 3. Scatter diagram $\dot{y} > 0$

What makes these figures interesting is that we can see that the right hand cap has essentially produced a perfect fold as suggested by Figure 4.

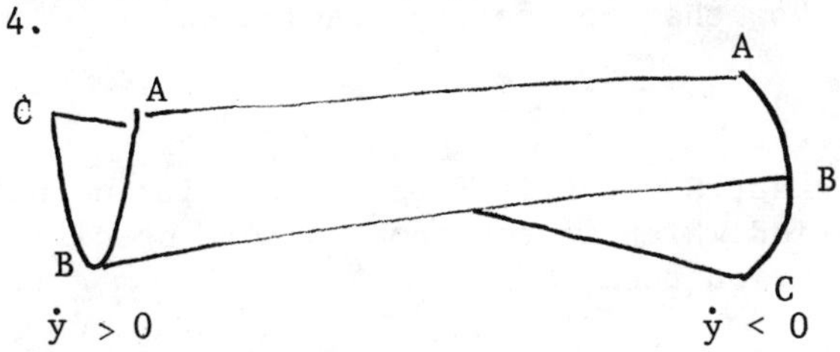

Figure 4. Folding the attractor in y > 1

Still more informative is the first return map associated with the set $y=1$, $\dot{y} < 0$. This is, of course, a map of a two dimensional set onto itself. However, because of the strong decay associated with the real eigenvalue, in the region $|y| > 1$ the trajectories which leave the regions $|y| > 1$ are very close to the plane defined by the eigenvectors associated with the (unstable) complex eigenvalues. This plane intersects the plan $y=1$ in a line and the (apparent) strange attractor lies close to this line. In Figure 5 we show the empirically determined first return map for the projections of the strange attractor on to the above line. A remarkable feature of this figure is the accuracy with which this relation is approximated by a function. Stated in another way, the flow generated by this differential equation is very well approximated by a semiflow which could, for example, be thought of as being generated by using the differential equations to propagate the motions from $y=1$, $\dot{y} > 0$ back to $y=1+\varepsilon$, $\dot{y} > 0$ and then use the above projection to complete the path back to $y=1$, $\dot{y} > 0$.

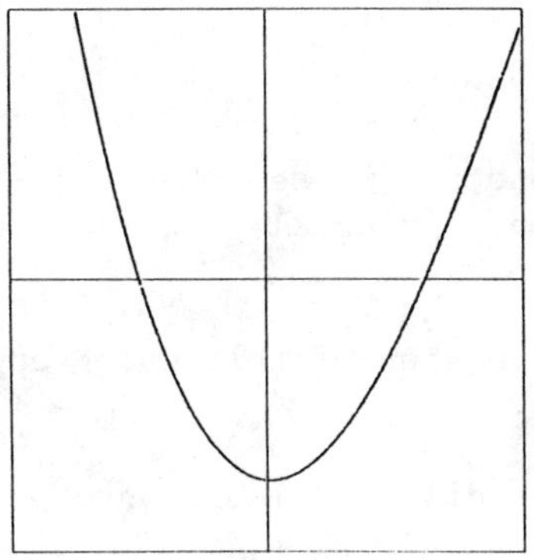

Figure 5. The first return map

3. Localization of Chaos. One of the more elegant applications of geometry in the study of dynamical systems involves the use of Fenchel's theorem to show that any periodic solution of

$$\dot{x} = f(x)$$

which is confined to a set S which is free of equilibrium points of f cannot have a period which is too short. More specifically, if T is a periodic solution then

$$\int_T |\partial f/\partial x f/\|f\|^2| \geq 2\pi$$

thus $T \geq 2\pi/\sup |\partial f/\partial x \cdot f|/\|f\|^2$. The periodic solutions cannot be too tightly confined unless there is an equilibrium point nearby. This result, resting as it does on a statement about the integrated curvature is a "second order" result.

There is an easily proven fact of this general type which applies to noninvertible onto maps of an interval into itself. In this context it says that one cannot confine a chaotic regime too tightly without having a large second derivative.

Lemma 1: If $f:[a,b]\to[a,b]$ is twice differentiable, onto, but not one-to-one, then

$$\varepsilon=\sup \left|\frac{\partial^2 f}{\partial x^2}\right| \geq \frac{1}{2|a-b|}$$

Proof: If f is onto then clearly

$$\int_a^b |\partial f/\partial x|dx \geq |a-b|$$

write

$$\frac{\partial f}{\partial x} = c + \int_0^x \frac{\partial^2 f}{\partial x^2} dm$$

Since f cannot be strictly monotone we see that $\varepsilon\,|a-b| \geq |c|$. Thus $|\partial f/\partial x| \leq 2\varepsilon|a-b|$ using the above integral inequality we see that the claim holds.

There is an n-dimensional version of this which applies to maps of $V\varepsilon R^n$ which goes as follows.

Lemma 2: If $f:V\to V$ is twice differentiable onto but not one-to-one then

$$\varepsilon \sup_V \left\|\frac{\partial^2 f}{\partial x^2}\right\| \geq \frac{(\text{diameter } V)^{n-1}}{2^n \text{ Vol } (V)}$$

We omit the proof which is somewhat tedious.

The relevance of this for differential equations may be argued as follows. If $\phi(t,x)$ denotes the solution at time t corresponding to $y(0)=x$ and

$$\dot{y} = f(y)$$

then we can compute the derivatives of ϕ with respect to x by solving certain differential equations. For example

$$\left.\frac{\partial \phi}{\partial x}\right]_x = \Phi(t)$$

where Φ is the matrix solution of

$$\dot{\Phi} = \left.\frac{\partial f}{\partial x}\right]_{\phi(t,x)} \Phi \quad ; \; \Phi(0) = I$$

Similarly, the 1,2 element of the fundamental solution of the extended linearization (see [5])

$$\frac{d}{dt}\begin{bmatrix} \delta \\ \delta^{[2]} \end{bmatrix} = \begin{bmatrix} \partial f/\partial x & \partial^2 f/\partial x^2 \\ 0 & (\partial f/\partial x)_{[2]} \end{bmatrix} \begin{bmatrix} \delta \\ \delta^{[2]} \end{bmatrix}$$

computes the second term in the Taylor series

$$\phi(t,x+\delta) = \phi(t,x) + \Phi(t)\delta + \Psi(t)\delta^{[2]} + \ldots$$

The point is that there exists an upper bound on the amount of "folding" that a differential equation can do and this upper bound involves the time available and bounds on $\partial^2 f/\partial x^2$ in the regions of interest.

Putting these pieces together, we see that given aprori bounds on the first two derivatives of f we can only strongly localize chaos by letting the first return time grow.

REFERENCES

[1] A.J. Lichtenberg and M.A. Lieberman, Regular and Stochastic Motion, Springer Verlag, New York, 1983.

[2] C.T. Sparrow, Chaos in a three-dimensional single loop feedback system with a piecewise linear feedback function, J. for Math. Analysis and Its Applications 83 (1981),pp. 275-291.

[3] R.W. Brockett, On conditions leading to chaos in feedback systems, Proceedings of the 21st IEEE Conference on Decision and Control Orlando, Florida, 2(1982), pp. 932-936.

[4] R.W. Brockett, Chaos and randomness in dynamical systems, Proceedings of the 22nd IEEE Conference on Decision and Control San Antonio, Texas, 1983 (with J. Loncaric).

[5] R.W. Brockett, Volterra series and geometric control theory, Automatica, Vol 12, pp. 167-176, 1976.

CHAOS IN DYNAMICAL SYSTEMS BY THE POINCARÉ-MELNIKOV-ARNOLD METHOD

JERROLD E. MARSDEN*

Abstract. Methods for proving the existence of chaos in the sense of Poincaré-Birkhoff-Smale horseshoes are presented. We shall concentrate on explicitly verifiable results that apply to specific examples such as the ordinary differential equations for a forced pendulum, and for superfluid ^{3}He and the partial differential equation describing the oscillations of a beam. Some discussion of the difficulties the method encounters near an elliptic fixed point is given.

1. An Introductory Example. Consider the equation for a forced pendulum

$$\ddot{\phi} + \sin\phi = \varepsilon \cos \omega t \tag{1.1}$$

where ω is a constant angular forcing frequency, and ε is a small parameter. For ε small but non-zero, (1.1) possess no analytic integrals of the motion. In fact, it possesses transversal intersecting stable and unstable manifolds (separatrices); that is, the Poincaré maps $P_{t_0}: \mathbb{R}^2 \to \mathbb{R}^2$ that advance solutions by one period $T = 2\pi/\omega$ starting at time t_0 possess transversal homoclinic points. This type of dynamic behavior has several consequences, besides precluding the existence of analytic integrals, that lead one to use the term 'chaotic'. For example, the equation (1.1) has infinitely many periodic solutions of arbitrarily high period. Also, using the shadowing lemma, one sees that given any bi-infinite sequence of zero and ones (for example, use the binary expansion of e or π), there exists a corresponding solution of (1.1) that successively crosses the plane $\phi = 0$ (the pendulum's vertically downward configuration) with $\dot{\phi} > 0$ corresponding to a zero and $\dot{\phi} < 0$ corresponding to a one. The origin of this chaos on an

*Research Group in Nonlinear Systems and Dynamics and Department of Mathematics, University of California, Berkeley, CA 94720. Research partially supported by DOE contract DE-AT03-82ER12097.

intuitive level lies in the motion of the pendulum near its unperturbed homoclinic orbit -- the orbit that does one revolution in infinite time. Near the top of its motion (where $\phi = \pm\pi$) small nudges from the forcing term can cause the pendulum to fall to the left or right in a temporally complex way.

The dynamical systems theory needed to justify all of the preceding, statements is now readily available in Smale [1967], Moser [1973] and Guckenheimer and Holmes [1983]. The key people responsible for the development of the basic theory are Poincare, Birkhoff and Smale. The idea of transversal intersecting separatrices comes from Poincaré's famous 1890 paper on the three body problem. His goal -- not quite achieved for reasons we shall comment on later -- was to prove the nonintegrability of the restricted three body problem and that various series expansions used up to that point diverged (he invented the theory of asymptotic expansions in the course of this work).

Although Poincaré had all the essential tools needed to prove that equations like (1.1) are not integrable (in the sense of having no analytic integrals) his interests lay with harder problems and he did not develop the easier basic theory very much. Important contributions were made by Melnikov [1963] and Arnold [1964] which leads to a very simple procedure for proving (1.1) is not integrable. The Poincare-Melnikov method was recently revived by Chirikov [1979], Holmes [1980] and Chow, Hale and Mallet-Paret [1980]. (For related work and more references and examples, see also Kozlov [1983].)

The procedure is as follows: rewrite (1.1) in abstract form as

$$\dot{x} = X_0(x) + \varepsilon X_1(x,t) \tag{1.2}$$

where $x \in \mathbb{R}^2$, X_0 is a Hamiltonian vector field with energy H_o, X_1 is periodic with period T and is Hamiltonian with energy H_1. Assume X_0 has a homoclinic orbit $\bar{x}(t)$ so $\bar{x}(t) \to x_0$, a hyperbolic saddle point, as $t \to \pm\infty$. Compute the "Melnikov function"

$$M(t_0) = \int_{-\infty}^{\infty} \{H_0, H_1\}(\bar{x}(t-t_0), t)\, dt \tag{1.3}$$

where $\{\ ,\ \}$ denotes the Poisson bracket. If $M(t_0)$ has simple zeros as a function of t_0, then (1.2) has transversal intersecting separatrices (in the sense of Poincare maps as mentioned above).

We shall give a proof of this result (essentially the one indicated by Arnold [1964] in §2. To apply it to equation (1.1) one proceeds as follows. Let $x = (\phi, \dot{\phi})$ so (1.1) becomes

$$\frac{d}{dt}\begin{pmatrix}\phi\\ \dot{\phi}\end{pmatrix} = \begin{pmatrix}\dot{\phi}\\ -\sin\phi\end{pmatrix} + \varepsilon\begin{pmatrix}0\\ \cos\omega t\end{pmatrix}. \tag{1.4}$$

The homoclinic orbits for $\varepsilon = 0$ are computed to be given by

$$\bar{x}(t) = \begin{pmatrix}\phi(t)\\ \dot{\phi}(t)\end{pmatrix} = \begin{pmatrix}\pm 2\tan^{-1}(\sinh t)\\ \pm 2\,\mathrm{sech}\, t\end{pmatrix} \tag{1.5}$$

and one has

$$\left.\begin{aligned} H_0(\phi,\dot{\phi}) &= \frac{1}{2}\dot{\phi}^2 - \cos\phi\\ H_1(\phi,\dot{\phi},t) &= \phi\cos\omega t\,. \end{aligned}\right\} \tag{1.6}$$

Hence (1.3) gives

$$\begin{aligned} M(t_0) &= \pm\int_{-\infty}^{\infty}\left(\frac{\partial H_0}{\partial\phi}\frac{\partial H_1}{\partial\dot{\phi}} - \frac{\partial H_0}{\partial\dot{\phi}}\frac{\partial H_1}{\partial\phi}\right)dt\\ &= \mp\int_{-\infty}^{\infty}\dot{\phi}\cos\omega t\,dt\\ &= \mp\int_{-\infty}^{\infty}[2\,\mathrm{sech}(t-t_0)\cos\omega t]\,dt\,. \end{aligned}$$

Changing variables and using the fact that sech is even and sin is odd, we get

$$M(t_0) = \mp 2\left(\int_{-\infty}^{\infty}\mathrm{sech}\,t\cos\omega t\,dt\right)\cos(\omega t_0)\,.$$

The integral is evaluated by residues:

$$M(t_0) = \mp 2\pi\,\mathrm{sech}\left(\frac{\pi\omega}{2}\right)\cos(\omega t_0) \tag{1.7}$$

which clearly has simple zeros.

2. A Proof of the Poincaré-Melnikov Theorem. There are two convenient ways of visualizing the dynamics of (1.2). One can introduce the Poincaré map $P_\varepsilon^s:\mathbb{R}^2 \to \mathbb{R}^2$, which is the time T map for (2.1) starting at time s. For $\varepsilon = 0$, the point x_0 and the homoclinic orbit are invariant under P_0^s, which is independent of s. The hyperbolic saddle x_0 persists as a nearby family of saddles x_ε for $\varepsilon > 0$, small, and we are interested in whether or not the stable and unstable manifolds of the point x_ε for the map P_ε^s intersect transversally (if this holds for one s, it holds for all s). If so, we say (1.2) admits horseshoes for $\varepsilon > 0$.

The second way to study (1.2) is to look directly at the suspended system on $\mathbb{R}^2 \times S^1$, where S^1 stands for the circle, elements of which are regarded as the T-periodic variable θ. Then (1.2) becomes the autonomous suspended system

$$\left.\begin{aligned} \dot{x} &= f_0(x) + \varepsilon f_1(x,\theta) \\ \dot{\theta} &= 1 \end{aligned}\right\} . \tag{2.1}$$

From this point of view, the curve

$$\gamma_0(t) = (x_0, t)$$

is a periodic orbit for (2.1), whose stable and unstable manifolds $W_0^s(\gamma_0)$ and $W_0^u(\gamma_0)$ are coincident. For $\varepsilon > 0$ the hyperbolic closed orbit γ_0 perturbs to a nearby hyperbolic closed orbit which has stable and unstable manifolds $W_\varepsilon^s(\gamma_\varepsilon)$ and $W_\varepsilon^u(\gamma_\varepsilon)$. If $W_\varepsilon^s(\gamma_\varepsilon)$ and $W_\varepsilon^u(\gamma_\varepsilon)$ intersect transversally, we again say that (1.2) admits horseshoes. These two definitions of admitting horseshoes are readily seen to be equivalent.

Poincaré-Melnikov Theorem. Define the Melnikov function by (1.3). Assume $M(t_0)$ has simple zeros as a T-periodic function of t_0. Then (1.2) has horseshoes.

Proof. In the suspended picture, we use the energy function H_0 to measure the first order movement of $W_\varepsilon^s(\gamma_\varepsilon)$ at $\bar{x}(0)$ at time t_0 as ε is varied. Note that points of $\bar{x}(t)$ are regular points for H_0 since H_0 is constant on $\bar{x}(t)$ and $\bar{x}(0)$ is not a fixed point. Thus, the values of H_0 give an accurate measure of the distance from the homoclinic orbit. If $(x_\varepsilon^s(t,t_0),t)$ is the curve on $W_\varepsilon^s(\gamma_\varepsilon)$ that is an integral curve of the suspended system (2.1) and has an initial condition $x^s(t_0,t_0)$ which is the perturbation of $W_0^s(\gamma_0)$ {the plane $t = t_0$} in the normal direction to the homoclinic orbit, then $H_0(x_\varepsilon^s(t_0,t_0))$ measures this normal distance. But

$$H_0(x_\varepsilon^s(T,t_0)) - H_0(x_\varepsilon^s(t_0,t_0)) = \int_{t_0}^{T} \frac{d}{dt} H_0(x_\varepsilon^s(t,t_0))\, dt \qquad (2.2)$$

From (2.2), we get

$$H_0(x_\varepsilon^s(T,t_0)) - H_0(x_\varepsilon^s(t_0,t_0) = \int_{t_0}^{T} \{H_0,H_0 + \varepsilon H_1\}(x^s(t,t_0,t)dt \qquad (2.3)$$

Since $x_\varepsilon^s(T,t_0)$ is ε-close to $\overline{x}(t-t_0)$ (uniformly as $T \to +\infty$), and $d(H_0 + \varepsilon H^1)(x^\varepsilon(t,t_0),t) \to 0$ exponentially as $t \to +\infty$, and $\{H_0,H_0\} = 0$, (2.3) becomes

$$H_0(x_\varepsilon^s(T,t_0)) - H_0(x_\varepsilon^s(t_0,t_0)) = \varepsilon \int_{t_0}^{T} \{H_0,H_1\}(\overline{x}(t-t_0,t)\, dt + O(\varepsilon^2). \qquad (2.4)$$

Similarly,

$$\begin{aligned} H_0(x_\varepsilon^u(t_0,t_0)) &- H_0(x_\varepsilon^u(-S,t_0)) \\ &= \varepsilon \int_{-S}^{t_0} \{H_0,H_1\}(\overline{x}(t-t_0),t)\, dt + O(\varepsilon^2) \end{aligned} \qquad (2.5)$$

Now $x_\varepsilon^s(T,t_0) \to \gamma_\varepsilon$, a periodic orbit for the perturbed system as $T \to +\infty$. Thus, we can choose T and S such that $H_0(x_\varepsilon^s(T,t_0)) - H_0(x_\varepsilon^u(-S,t_0)) \to 0$ as $T,S \to \infty$. Thus, adding (2.4) and (2.5), and letting $T,S \to \infty$, we get

$$H_0(x_\varepsilon^u(t_0,t_0)) - H_0(x_\varepsilon^s(t_0,t_0)) = \varepsilon \int_{-\infty}^{\infty} \{H_0,H_1\}(\bar{x}(t-t_0),t)\, dt + O(\varepsilon^2) \tag{2.6}$$

It follows that if $M(t_0)$ has a simple zero in time t_0, then $x_\varepsilon^u(t_0,t_0)$ has $x_\varepsilon^s(t_0,t_0)$ must intersect transversally near the point $\bar{x}(0)$ at time t_0.

Remark. Since $dH_0 \to 0$ exponentially at the saddle points, the integrals involved in this criterion are automatically convergent.

3. An Extension to Include Damping. There are a number of extensions and applications of this technique that have been developed, some of which we describe here and in the next few sections. The literature in this area is growing very quickly and we make no claim to be comprehensive (the reader can track down many additional references by consulting the references cited).

If in (1.2), X_0 is Hamiltonian but X_1 is not, the same conclusion holds if (1.3) is replaced by

$$M(t_0) = \int_{-\infty}^{\infty} (X_0 \times X_1)(\bar{x}(t-t_0),t)\, dt \tag{3.1}$$

where $X_0 \times X_1$ is the (scalar) cross product for planar vector fields. In fact, X_0 need not even be Hamiltonian if a volume expansion factor is inserted.

For example, this applies to the forced damped Duffing equation

$$\ddot{u} - \beta u + \alpha u^3 = \varepsilon(\gamma \cos \omega t - \delta \dot{u}) \tag{3.2}$$

Here the homoclinic orbits are given by

$$\bar{u}(t) = \pm \sqrt{\frac{\beta}{\alpha}}\, \operatorname{sech}(\sqrt{\beta} t) \tag{3.3}$$

and (3.1) becomes, after a residue calculation,

$$M(t_0) = 2\gamma\, \pi\omega \sqrt{\frac{2}{\alpha}}\, \operatorname{sech}\left(\frac{\pi\omega}{2\sqrt{\beta}}\right) \sin(\omega t_0) + \frac{4\delta\beta^{3/2}}{3\alpha} \tag{3.4}$$

so one has simple zeros and hence chaos of the horseshoe type if

$$\frac{\gamma}{\delta} > \frac{\sqrt{2}\,\beta^{3/2}}{3\omega\sqrt{\alpha}} \cosh\left(\frac{\pi\omega}{2\sqrt{\beta}}\right) \tag{3.5}$$

and ε is small.

Another interesting example, due to Montgomery [1984] concerns the equations for superfluid ^{3}He. These are the Leggett equations and we shall confine ourselves to the A phase for simplicity (see Montgomery's paper for additional results). The equations are

$$\begin{aligned} \dot{s} &= -\frac{1}{2}\left(\frac{\chi\Omega^2}{\gamma^2}\right)\sin 2\theta \\ \dot{\theta} &= \left(\frac{\gamma^2}{\chi}\right)s - \varepsilon(\gamma B \sin \omega t + \frac{1}{2}\Gamma \sin 2\theta) \end{aligned} \tag{3.6}$$

Here s is the spin, θ the angle describing the order parameter and $\gamma, \chi, \ldots$ are physical constants. The homoclinic orbits for $\varepsilon = 0$ are given by

$$\begin{aligned} \bar{\theta}_{\pm} &= 2\tan^{-1}(e^{\pm\Omega t}) - \pi/2 \\ \bar{s}_{\pm} &= \pm 2\,\frac{\Omega e^{\pm 2\Omega t}}{1 + e^{\pm 2\Omega t}} \end{aligned} \tag{3.7}$$

One calculates using (3.6) and (3.7) in (3.1) that

$$M_{\pm}(t_0) = \mp\frac{\pi\chi\omega B}{8\gamma}\operatorname{sech}\left(\frac{\omega\pi}{2\Omega}\right)\cos\omega t - \frac{2}{3}\frac{\chi}{\gamma^2}\Omega\Gamma \tag{3.8}$$

so that (3.6) has chaos in the sense of horseshoes if

$$\frac{\gamma B}{\Gamma} > \frac{16}{3\pi}\frac{\Omega}{\omega}\cosh\left(\frac{\pi\omega}{2\Omega}\right) \tag{3.9}$$

and if ε is small.

4. <u>An Extension to PDE's.</u> There is a version of the Poincaré-Melnikov theorem applicable to PDE's that is due to Holmes and Marsden [1981]. One basically still uses the formula (3.1) where

$X_0 \times X_1$ now is replaced by the symplectic pairing between X_0 and X_1. However, there are two new difficulties in addition to standard technical analytic problems that arise with PDE's. The first is that there is a serious problem with resonances. These can be dealt with using the aid of damping -- the undamped case would need an infinite dimensional version of Arnold diffusion -- see §6 below. Secondly, the problem is not reducible to two dimensions; the horseshoe involves all the modes. Indeed, the higher modes do seem to be involved in the physical buckling processes for the beam model discussed next.

A PDE model for a buckled forced beam is

$$\ddot{w} + w'''' + \Gamma^1 w'' - \kappa\left(\int_0^1 [w']^2 dz\right) w'' = \varepsilon(f \cos \omega t - \delta\dot{w}) \tag{4.1}$$

where $w(z,t)$ $0 \leq z \leq 1$ describes the deflection of the beam, $\dot{} = \partial/\partial t$, $' = \partial/\partial z$ and $\Gamma, \kappa, \ldots$ are physical constants. For this case, the theory shows that if

(a) $\pi^2 < \Gamma < 4\rho^3$ (first mode is buckled)

(b) $j^2\pi^2(j^2\pi^2 - \Gamma^1) \neq \omega^2$, $j = 2,3, \ldots$ (resonance condition)

(c) $\frac{f}{\delta} > \frac{\pi(\Gamma-\pi^2)}{2\omega\sqrt{\kappa}} \cosh\left(\frac{\omega}{2\sqrt{\Gamma-\omega^2}}\right)$ (transversal zeros for $M(t_0)$)

(d) $\delta > 0$

and ε is small, then (4.1) has horseshoes.

Experiments of F. Moon at Cornell which show chaos in a forced buckled beam provided the motivation which led to the study of (4.1).

This kind of result has recently been used by Slemrod and Marsden [1983] for a study of chaos in a van der Wall's fluid (see Slemrod's lecture in these proceedings) and by Roos, Birnir and Morrison for soliton equations. For example, in the damped, forced Sine-Gordon equation one has chaotic transitions between breathers and kink-antikink pairs and in the Benjamin-Ono equation one can have chaotic transitions between solutions with different numbers of poles.

5. Autonomous Hamiltonian Systems. For Hamiltonian systems with two degrees of freedom, Holmes and Marsden [1982a] show how the Melnikov method may be used to prove the existence of horseshoes on energy surfaces in two degree of freedom nearly integrable systems. The class of systems studied have a Hamiltonian of the form

$$H(q,p,\theta,I) = F(q,p) + G(I) + \varepsilon H_1(q,p,\theta,I) + O(\varepsilon^2) \tag{5.1}$$

where (θ,I) are action angle coordinates for the oscillator G; $G(0) = 0$, $G' > 0$. It is assumed that F has a homoclinic orbit $\bar{x}(t) = (\bar{q}(t), \bar{p}(t))$ and that

$$M(t_0) = \int_{-\infty}^{\infty} \{F,H_1\}\, dt \tag{5.2}$$

(the integral taken along $(\bar{x}(t-t_0), \Omega t, I))$ has simple zeros. Then (5.1) has horseshoes on energy surfaces near the surface corresponding to the homoclinic orbit and small I; the horseshoes are taken relative to a Poincaré map strobed to the oscillator G. Holmes and Marsden 1982a also studies the effect of positive and negative damping. These results are related to that in §2 since one can often reduce a two degree of freedom Hamiltonian system to a one degree of freedom forced system.

For some systems in which the variables do not split as in 5.1, such as a nearly symmetric heavy top, one needs to exploit a symmetry of the system and this complicates the situation to some extent. The general theory for this is given in Holmes and Marsden [1983] and was applied to show the existence of horseshoes in the nearly symmetric heavy top; see also some closely related results of Ziglin [1980a].

This theory has been used, for example by Koiller and coworkers in a number of recent reprints on vortex dynamics (Koiller and Pinto de Carvalho [1983] seems to be the first to give a correct proof of the non-integrability of the restricted four vortex problem -- see §7 below). There have also been recent applications to the dynamics of general relativity showing the existence of horsehows in Bianchi IX models. See also Krishnaprasad [1983] for interesting applications to dual-spin spacecraft.

6. Arnold Diffusion. Arnold [1964] extended the Poincaré-Melnikov theory to systems with several degrees of freedom. In this case the transverse homoclinic manifolds are based on KAM tori and allow the possibility of chaotic drift from one torus to another. This drift, now known as Arnold diffusion is a basic ingredient in the study of chaos in Hamiltonian systems (see for instance, Chirikov [1979] and Lichtenberg and Lieberman [1983] and references therein). Instead of a single Melnikov function, one now has a Melnikov vector given schematically by

$$\vec{M} = \begin{pmatrix} \int_{-\infty}^{\infty} \{H_0, H_1\}\, dt \\ \\ \int_{-\infty}^{\infty} \{I_k, H_1\}\, dt \end{pmatrix} \tag{6.1}$$

where I_k are integrals for the unperturbed (completely integrable) system and where $\vec{M}$ now depends on t_0 and on angles conjugate to $I_1, \ldots, I_n$. One now requires $\vec{M}$ to have transversal zeros in the vector sense. This result was given by Arnold for forced systems and was extended to the autonomous case by Holmes and Marsden [1982b]*, [1983].

These results apply to systems such as a pendulum coupled to several oscillators and the many vortex problem. It has also been used in power systems by Salam, Marsden and Varaiya [1984], building on the horseshoe case treated by Kopell and Washburn [1982]. See also the work of Salam and Sastry reported in these proceedings.

There have been a number of other directions of research on these techniques. For example, Grundler [1981] developed a multidimensional version applicable to the spherical pendulum and Greenspan and Holmes [1983] showed how it can be used to study subharmonic bifurcations.

7. Exponentially Small Melnikov Functions. There is a serious difficulty that arises when one uses the Melnikov method near an elliptic fixed point in a Hamiltonian system. The difficulty is closely related to the difficulty Poincaré encountered in trying to prove nonintegrability and the divergence of series expansions that occur in the restricted 3 body problem. Near elliptic points, one sees homoclinic orbits in normal forms and after a temporal rescaling leads to a loss of analyticity and a rapidly oscillatory perturbation that is modelled by the following variation of (1.1):

$$\ddot{\phi} + \sin\phi = \varepsilon \cos\left(\frac{\omega t}{\varepsilon}\right) \tag{7.1}$$

If one just blindly computes $M(t_0)$ one finds from (1.7),

---

* As was pointed out by F.A. Salam and C. Robinson, one needs to interpret the integrals appearing here with care and correctly adjust the phases of orbits asymptotic to the tori.

$$M(t_0,\varepsilon) = \mp 2\pi \operatorname{sech}\left(\frac{\pi\omega}{2\varepsilon}\right) \cos\left(\frac{\omega t_0}{\varepsilon}\right) \tag{7.2}$$

while this has simple zeros, the proof of the Poincaré-Melnikov theorem is no longer valid since $M(t_0,\varepsilon)$ is now of order $e^{-\pi/2\varepsilon}$ and the error analysis in the proof only gives errors of order ε^2. In fact no expansion in powers of ε can detect exponentially small terms like $e^{-\pi/2\varepsilon}$. (This is the sort of difficulty that seems to occur in the paper of Ziglin [1980b] on the four vortex problem; see also Sanders [1982].)

Recent work of Holmes, Marsden and Scheurle aims to show that indeed (7.1) has horseshoes for small ε. The idea is to expand expressions for the stable and unstable manifolds in a Perron type series whose terms are of order $^k e^{-\pi/2\varepsilon}$. To do so, the extension of the system to compelx time plays a crucial role.

One can hope that if such results for (7.1) can really be proven, then it may be possible to return to Poincaré's 1890 work and complete the arguments he left unfinished.

References

[1] F.H. ABDEL-SALAM, J.E. MARSDEN and P.P. Varaiya, Arnold Diffusion in the Swing Equations of a Power System, (1983b), Trans. IEEE (to appear).

[2] V. ARNOLD, Instability of Dynamical Systems with Several Degrees of Freedom, Dokl. Akad. Nauk. SSSR. 156(1964) pp. 9-12.

[3] B.V. CHIRIKOV, A Universal Instability of Many-Dimensional Oscillator Systems, Physics Reports, 52(1979), pp. 265-379.

[4] S.N. CHOW, J.K. HALE, and J. MALLET-PARET, An Example of Bifurcation to Homoclinic Orbits, J. Diff. Eqns., 37(1980), pp. 351-373.

[5] B. GREENSPAN and P. HOLMES, Subharmonic Bifurcations and Melnikov's Method, (1983), preprint.

[6] J. GRUNDLER, Thesis, University of North Carolina (1981).

[7] J. GUCKENHEIMER and P. HOLMES, Nonlinear Oscillations, Dynamical Systems, and Bifurcation of Vector Fields, Springer, Applied Math. Sciences, Vol. 42, 1983.

[8] P. HOLMES, Averaging and Chaotic Motions in Forced Oscillations SIAM J. on Appl. Math. 38, 68-80 and 40(1980), pp. 167-168.

[9] P.J. HOLMES and J.E. MARSDEN, A Partial Differential Equation with Infinitely Many Periodic Orbits: Chaotic Oscillations of a Forced Beam, Arch. Rat. Mech. Anal. 76(1981), pp. 135-166.

[10] P.J. HOLMES and P.E. MARSDEN, Horseshoes in Perturbations of Hamiltonian Systems with Two Degrees fo Freedom, Comm. Math. Phys. 82(1982a), pp. 523-544.

[11] P.J. HOLMES and J.E. MARSDEN, Melnikov's Method and Arnold Diffusion for Perturbations of Integrable Hamiltonian Systems, J. Math. Phys. 23(1982b), pp. 669-675.

[12] P. HOLMES and J. MARSDEN, Horseshoes and Arnold Diffusion for Hamiltonian Systems on Lie Groups, Ind. Univ. Math. J., 32(1983), pp. 273-310.

[13] J. KOILLER, A dynamical System with a Wild Horseshoe, 1983 (preprint).

[14] J. KOILLER and S. Pinto de Carvalho, Nonintegrability of a Restricted Problem of Four Point Vortices 1983, (preprint).

[15] N. KOPELL and R.B. WASHBURN, Chaotic Motions in the Two-Degree-of-Freedom Swing Equations, IEEE Trans. Circuits and Systems 29(1982), pp. 738-746.

[16] S. KOZLOV, Nonintegrability in Hamiltonian Systems, Russ. Math Surveys, 38(1983), pp. 1-76.

[17] P.S. KRISHNAPRASAD, Lie-Poisson Structures and Dual-Spin Spacecraft, 1983 (preprint).

[18] A.J. LICHTENBERG and M.A. LIEBERMANN, Regular and Stochastic Motion, Springer, Applied Math. Sciences, Vol. 38, 1983.

[19] V.K. MELNIKOV, On the Stability of the Center for Time Periodic Perturbations, Trans. Moscow Math. Soc. 12(1963), pp. 1-57.

[20] R. MONTGOMERY, Chaos in the Leggett Equation for Superfluid ^{3}He, 1984 (preprint).

[21] J. MOSER, Stable and Random Motions in Dynamical Systems, Annals of Math. Studies, Princeton University Press, 1973.

[22] J. SANDERS, Melnikov's Method and Averaging, Celest. Mech. 28 28(1982), pp. 171-181.

[23] M. SLEMROD and J. MARSDEN, Temporal and Spatial Chaos in a van der Waals Fluid due to Periodic Thermal Fluctuations, Adv. in Appl. Math, 1983 (to appear).

[24] S. SMALE, Differentiable Dynamical Systems, Bull. Am. Math. Soc., 73(1967), pp. 747-817.

[25] S.L. ZIGLIN, Decomposition of Separatrices, Branching of Solutions and Nonexistence of an Integral in the Dynamics of a Rigid Body, Trans. Moscow Math. Soc., 41(1980a), p. 287.

[26] S.L. ZIGLIN, Nonintegrability of a Problem on the Motion of Four Point Vortices, Sov. Math. Dokl., 21(1980b), pp. 296-299.

[27] S.L. ZIGLIN, Branching of Solutions and Nonexistence of Integrals in Hamiltonian Systems. Doklady Akad Nauk. SSSR 257(1981a), pp. 26-29.

[28] S.L. ZIGLIN, Self-intersection of the Complex Separatrices and the Nonexistence of the Integrals in the Hamiltonian Systems with 1-1/2 Degrees of Freedom, Prikl. Math. Mek. 45, 564-566; trans. as J. Appl. Math. Mech., 45(1981b), pp. 411-413

SPATIAL CHAOS IN FIRST ORDER PHASE TRANSITIONS OF A VAN DER WAALS FLUID

M. SLEMROD*

Abstract. This note discusses the application of the Holmes, Marsden Mel'nikov approach to chaos to the problem of the equilibrium distribution of a van der Waals fluid undergoing spatially periodic thermal variations.

0. Introduction In a recent paper [1] J. Marsden and I have considered the implications of the Holmes-Marsden Mel'nikov approach to chaos for both dynamic and equilibrium configuration of a van der Waals fluid. We showed in that paper that fluctuctions in temperature which are (i) temporally periodic can induce temporal chaos while temperature fluctuations that are (ii) spatially periodic can induce spatial chaos. In this note, I review the basic ideas of the route to spatial chaos for the equilibrium fluid distribution.

The rest of the paper is divided into three sections. Section 1 reviews the classical van der Waals theory of first order phase transitions. Section 2 provides a review of the basic Mel'nikov function approach to chaos as formulated by Holmes and Marsden. Finally, Section 3 applies the Mel'nikov function technique to the thermally driven van der Waals fluid.

1. The classical isothermal first order phase transition. Consider the following experiment described in [2]. Water vapor is put in a large container at 15°C. This is sketched in Fig. 1. We now start to move a piston into the container. As we move the piston the density of the vapor increases and so the specific volume $V = (\text{density})^{-1}$ decreases. For a while the pressure p inside the container increases. This depicted as the E_1-E_2 curve in Fig. 2. When we reach E_2 liquid drops of water start to form and as we bring the

---

*Department of Mathematical Sciences, Rensselaer Polytechnic Institute, Troy, N. Y. 12180. This research was sponsored in part by the Air Force Office of Scientific Research, Air Force Systems Command, USAF, under Contract/Grant No. AFOSR-81-0172. The United States Government is authorized to reproduce and distribute reprints for government purposes not withstanding any copyright herein.

piston down the pressure stays constant until we reach F_1. Here all the vapor has condensed to liquid at which point the pressure again rises with decreasing specific volume. This is shown on the F_1-F_2 curve. Notice the departure from the ideal gas law $p = RTV^{-1}$, R = positive constant, T = absolute temperature, which is sketched in Fig. 3.

To account for the possible co-existence of liquid and vapor phases found in many fluids, van der Waals in 1873 introduced the constitutive relation that bears his name.

(1) $$p = \frac{RT}{V-b} - \frac{a}{V^2} \qquad \begin{array}{l} 0 < b < V < \infty, \\ R,\ a,\ b \text{ pos. constants.} \end{array}$$

In (1) the first term represents a correction to the ideal gas law based on the idea that a fluid cannot be compressed beyond some critical molecular size. The second term accounts for inter-particle attractiveness. Notice that when V is small the second term becomes dominant, i.e., precisely when the fluid is in the liquid phase the inter-particle attractions which are neglected in the ideal gas law are now accounted for. A graph of a van der Waals isotherm for $T < T_{critical} = 8a\,(27bR)^{-1}$ is given in Fig. 4.

From Fig. 4 we see that for a certain range of pressures, $p_\alpha < p < p_\beta$, the van der Waals fluid can simultaneously exist in liquid $(b < V < \alpha)$ and vapor phases $(V > \beta)$. However, this is not what we observed in our hypothetical experiment for water. Specifically, our experiment noted a <u>preferred pressure</u> $p_{eq.}$ as given on the F_2-F_1 line in Fig. 2 for co-existence of liquid and vapor phases. One way to reconcile the van der Waals relation (1) and the experimental result sketched in Fig. 2. was given by Maxwell using thermodynamics and this is the presentation found in standard texts [3], [4]. However in 1893 [5], [6] van der Waals, himself, presented a resolution of this problem in a different manner which is particularly enlightening for the purposes of this note. Van der Waals' idea has come down to us via subsequent refinements [7] and statistical derivations [8]. However, in this note I'll present a simplified one dimensional theory which is sufficient for our purposes.

Consider one dimensional fluid flow along the x-axis in a tube of unit cross section. According to van der Waals [6] the usual thermo-elastic constitutive relation for the Cauchy stress $\tau = -p(V,T)$ should be replaced by the relationship

(2) $$\tau = p(V,T) - AV''$$

where $' = \frac{d}{dx}$. For simplicity, we take A to be a positive constant. The term $-AV''$ represents the addition of interfacial capillarity to the Cauchy stress.

In the absense of body forces the balance of linear momentum for equilibrium distribution of the fluid is just

$$\tau' = 0$$

and hence (2) implies

$$AV'' + p(V,T) = B, \qquad -\infty < x < \infty, \tag{3}$$

where B is a constant. For solutions V(x) of (3) which possess limits as $|x| \to \infty$, B represents the pressure of the "ends" of the tube.

For co-existence of liquid and vapor phases at isothermal ($T = T_0 < T_{crit.}$) equilibrium we require

$$V(x) \to V_m \text{ (liquid phase, } b < V_m < \alpha), \ x \to \infty,$$

$$V(x) \to V_M \text{ (vapor phase, } \beta < V_M), \ x \to -\infty.$$

In this case (3) becomes

$$AV'' + p(V,T_0) = B \tag{4}$$

and $B = p(V_m,T_0) = p(V_M,T_0)$. Multiply (4) by $V'(x)$ so that

$$\left(\frac{AV'(x)^2}{2}\right)' + \left(p(V,T_0) - p(V_m,T_0)\right)V'(x) = 0 \tag{5}$$

Equation (5) can now be used to represent V(x) as a quadrature. Furthermore, we immediately see that integration of (5) from $-\infty$ to ∞ shows

$$\int_{-\infty}^{\infty} \left(p(V,T_0) - p(V_m,T_0)\right)V'(x)dx = 0 \tag{6}$$

or with a change of variable of integration

$$\int_{Vm}^{V_M} \left(p(V,T_0) - p(V_m,T_0)\right)dV = 0. \tag{7}$$

Equation (7) says that for co-existence of liquid and vapor phase V_M, V_m must be related by (7), i.e., the Maxwell equal area rule shown in Fig. 4. This is the same rule formulated by Maxwell by recourse to the first law of thermodynamics. (E.g., see [3], [4].)

According to these classical arguments the experimental result of Fig. 2 now is understood. The co-existence of phases described by F_2-F_1 takes place when the end points are the values V_m, V_M given by the Maxwell rule (7).

Of course (4) admits solutions which do not possess co-existing phase limits as $|x| \to \infty$. In Section 3, we study the structure stability of solutions of (4) when T_0 is perturbed by a small periodic variation. First, however, we recall the Holmes-Marsden-Mel'nikov theory of "chaos".

2. Holmes & Marsden, Mel'nikov theory. Consider the perturbed Hamiltonian system in R^2 given by

(8) $$\dot{z}(t) = f_0(z(t)) + \varepsilon f_1(z(t),t)$$

where $z \in R^2$, f_0: $R^2 \to R^2$, f_1: $R^2 \times [0, \frac{2\pi}{\omega}] \to R^2$. We assume

(i) (8, $\varepsilon = 0$) is Hamiltonian;

(ii) $f_1(z,\bullet)$ is $\frac{2\pi}{\omega}$ periodic;

(iii) $f_0(p_0) = 0$;

(iv) For $\varepsilon = 0$ p_0 is a homoclinic saddle point in that (8) possesses a homoclinic orbit $q^\circ(t)$ connecting p_0 to itself.

Define the Mel'nikov function

$$M(t_0) = \int_{-\infty}^{\infty} f_0(q^\circ(t-t_0)) \wedge f_1(q^\circ(t-t_0),t)dt.$$

Lemma 1. ([9], p. 186). Under the above assumptions for ε sufficiently small (8) possesses a unique hyperbolic periodic orbit $\gamma_\varepsilon^\circ(t) = p_0 + O(\varepsilon)$. Correspondingly the Poincare map $P_\varepsilon^{t_0}$ has a unique hyperbolic saddle point $p_\varepsilon^{t_0} = p_o + O(\varepsilon)$.

Theorem 1. ([9], p. 188). If $M(t_0)$ has simple zeros and is independent of ε, then for $\varepsilon > 0$ sufficiently small the stable and unstable manifolds associated with $p_\varepsilon^{t_0}$ intersect transversely. Furthermore, some iterate of the Poincare map has an invariant hyperbolic set: a Smale horseshoe.

Remark 1. A similar result holds if p_0 is connected via a heteroclinic orbit to another saddle point p_1 for (8) with $\varepsilon = 0$. (See Holmes and Marsden [10].)

3. Spatial chaos in the equilibrium configuration of a van der Waals fluid. We consider three cases.

Case 1. The constant B is such that $p(V_M,T_0) < B < p(\beta,T_0)$. In this case the V-V' phase portraits of (4) is shown in Fig. 5.

Case 2. The constant B is such that $p(\alpha,T_0) < B < p(V_M,T_0)$. In this case the V-V' phase portraits of (4) are shown in Fig. 6.

Case 3. The constant B is such that $B = p(V_M,T_0) = p(V_m,T_0)$. In this case the V-V' phase portrait is shown in Fig. 7.

Now we note that for a small thermal perturbation away from T_0 of the form

$$T(x) = T_0 + \varepsilon \cos \omega x$$

the van der Waals relation (1) becomes

$$p(V,T) = p(V,T_0) + \varepsilon \frac{R \cos \omega x}{V-b}$$

and (3) takes the form

$$AV'' + p(V,T_0) + \varepsilon \frac{R \cos \omega x}{V-b} = B \tag{9}$$

where p is given by (1). But (9) is a perturbed Hamiltonian system of the type described in Section 2. The Mel'nikov function in all three cases is of the form

$$M(x_0) = \int_{-\infty}^{\infty} \frac{Y'(x-x_0)\cos \omega x \, dx}{(Y(x-x_0)-b)}$$

where Y(x) denotes either a homoclinic or heteroclinic orbit. From an elementary trigonometric identity we find

$$M(x_0) = RL \cos \omega x_0 + RN \sin \omega x_0$$

where

$$L = \int_{-\infty}^{\infty} \frac{Y'(x)\cos \omega x \, dx}{(Y(x)-b)} \quad , \qquad N = \int_{-\infty}^{\infty} \frac{Y'(x)\sin \omega x \, dx}{(Y(x)-b)}$$

If $L \neq 0$, $N \neq 0$, $M(x_0)$ has a simple zero at $\omega x_0 = \arctan(L/N)$. If $N = 0$, $L \neq 0$, $M(x_0)$ has a simple zero at $\omega x_0 = (2m+1)\pi/2$, m any integer. If $L = 0$, $N \neq 0$, $M(x_0)$ has a simple zero at $\omega x_0 = m\pi$, m any integer. Hence, $M(x_0)$ possesses a simple zero in all three cases. If we apply Thm. 1 and Remark 1 of Section 2, we can state the following result.

Theorem 2. Consider the equilibrium configuration of a van der Waals fluid undergoing thermal variations $T(x) = T_0 + \varepsilon \cos \omega x$ $T_0 < T_{crit}$. Then for an applied stress B, $p(\alpha, T_0) < B < p(\beta, T_0)$ on equilibrium configuration will exhibit spatial chaos in the sense of possessing horseshoes for ε sufficiently small.

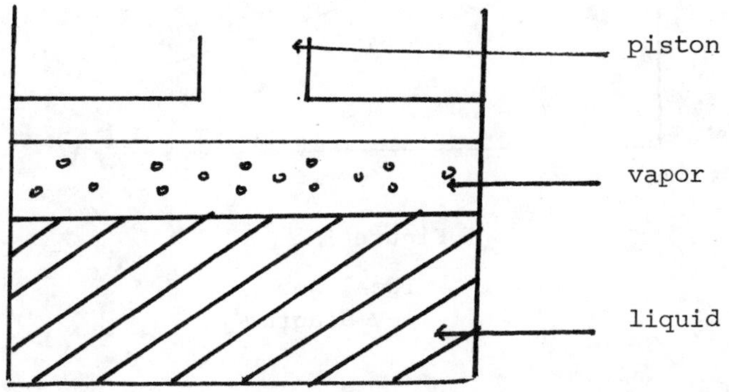

Figure 1.

Water vapor in large container

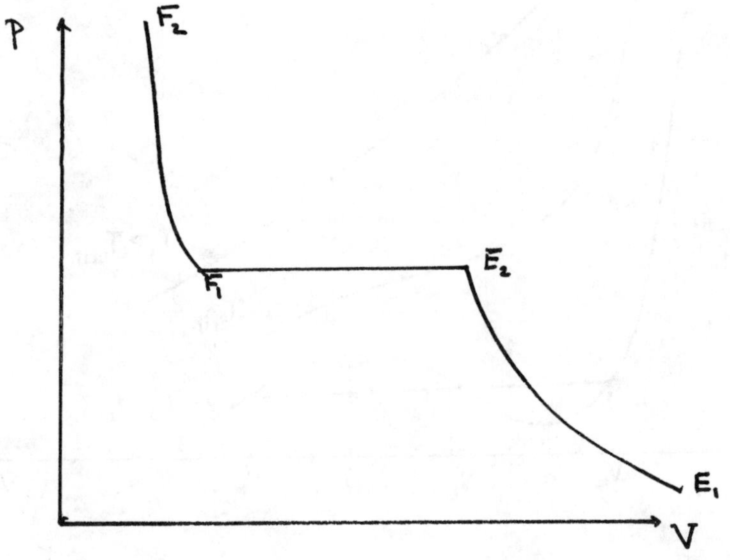

Figure 2.

Water 15^0C p-V diagram

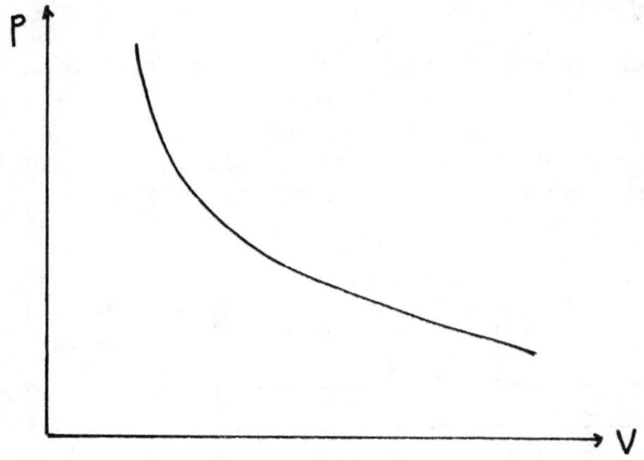

Figure 3.
Ideal gas
p-V diagram

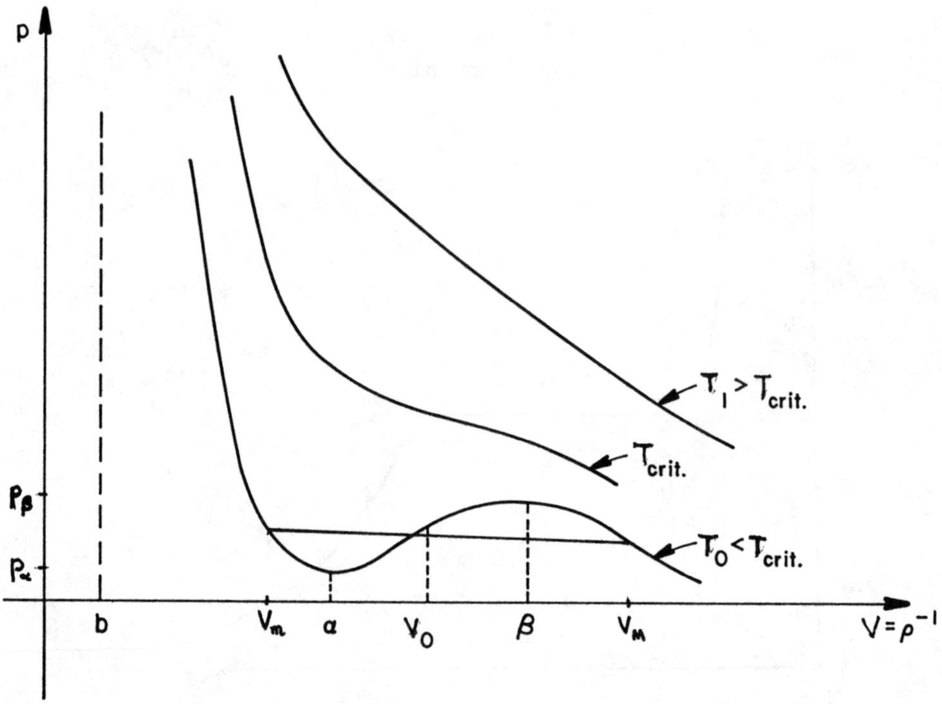

Figure 4.
van der Waals isotherms
and the Maxwell construction

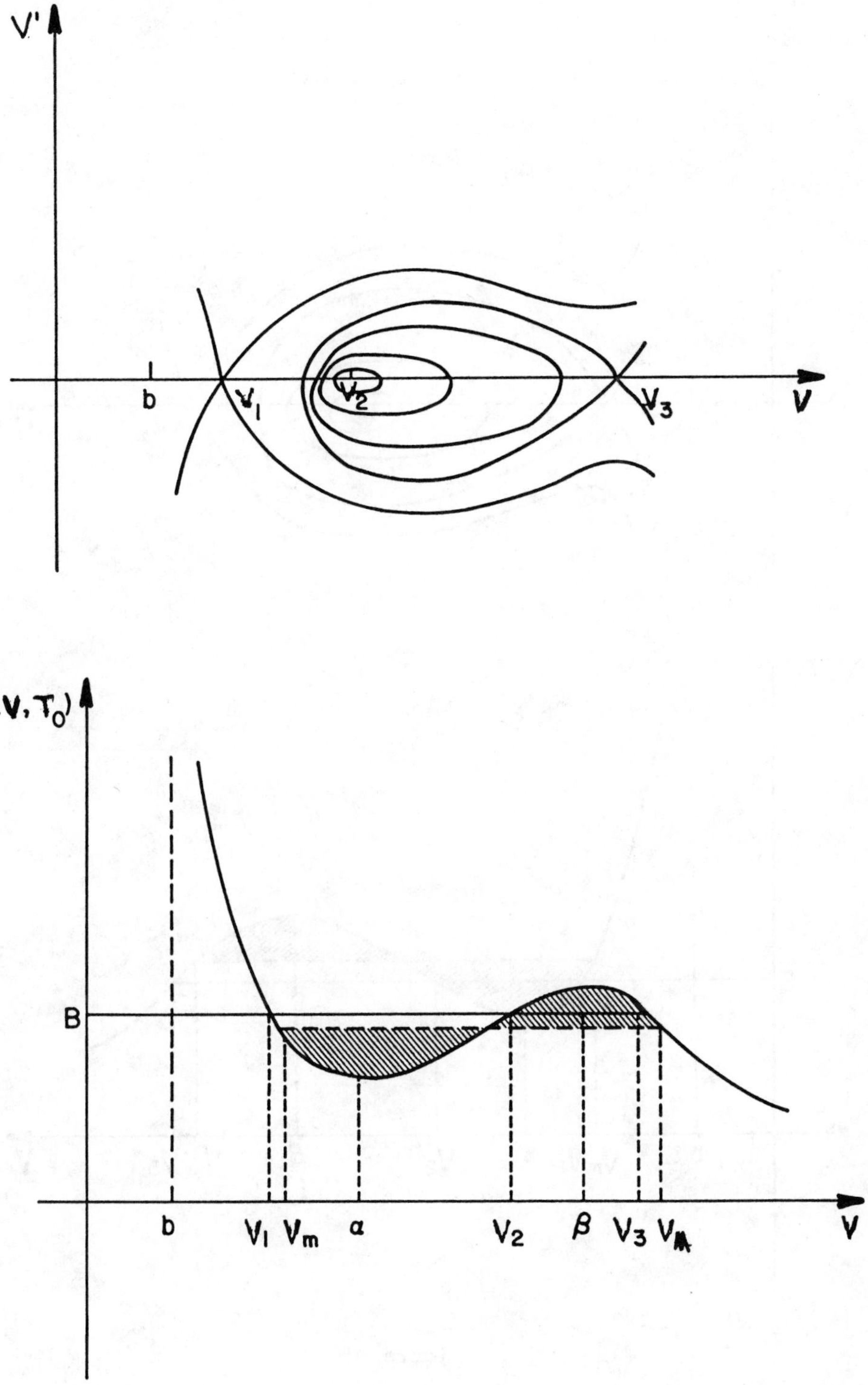

Figure 5

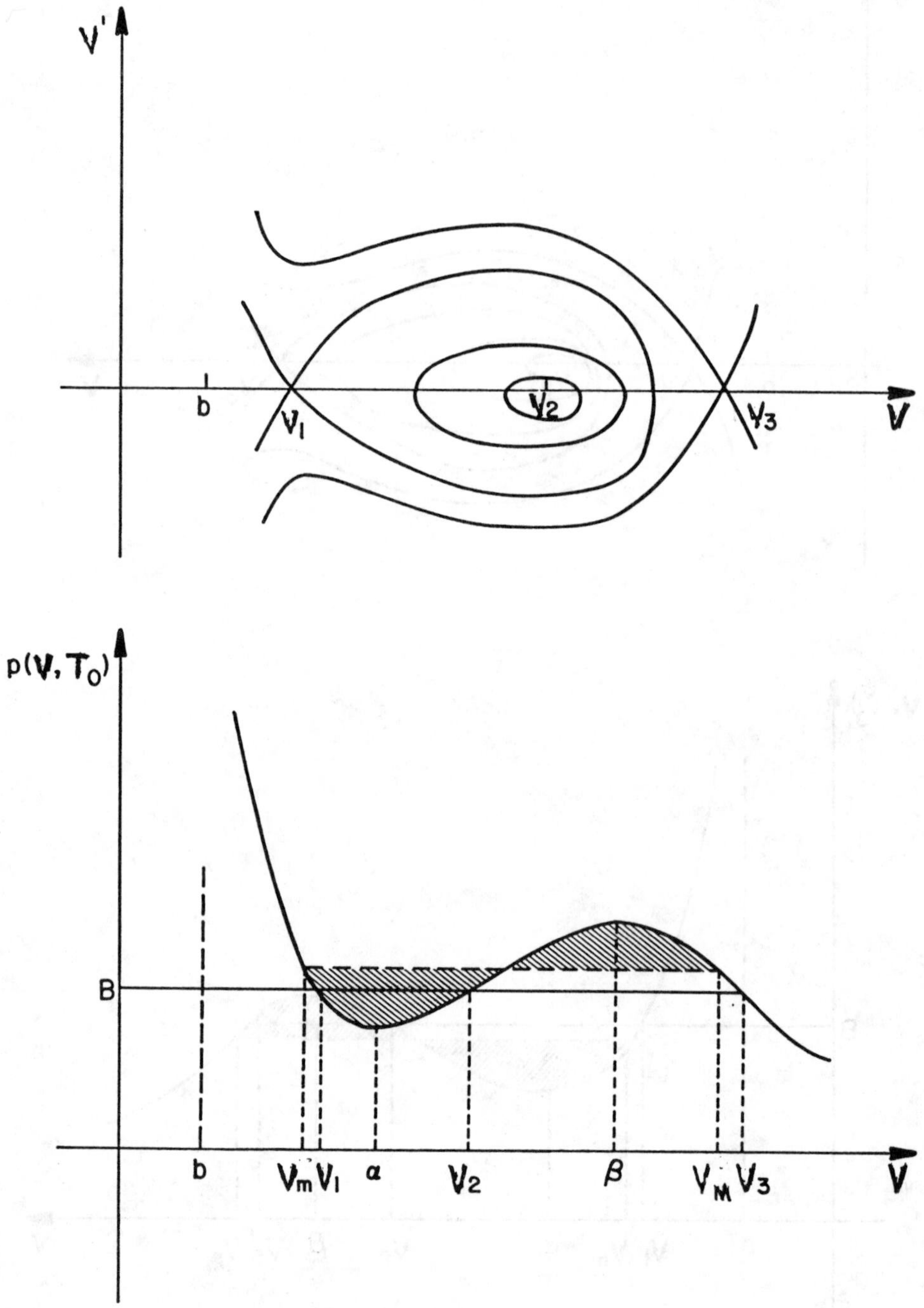

Figure 6.

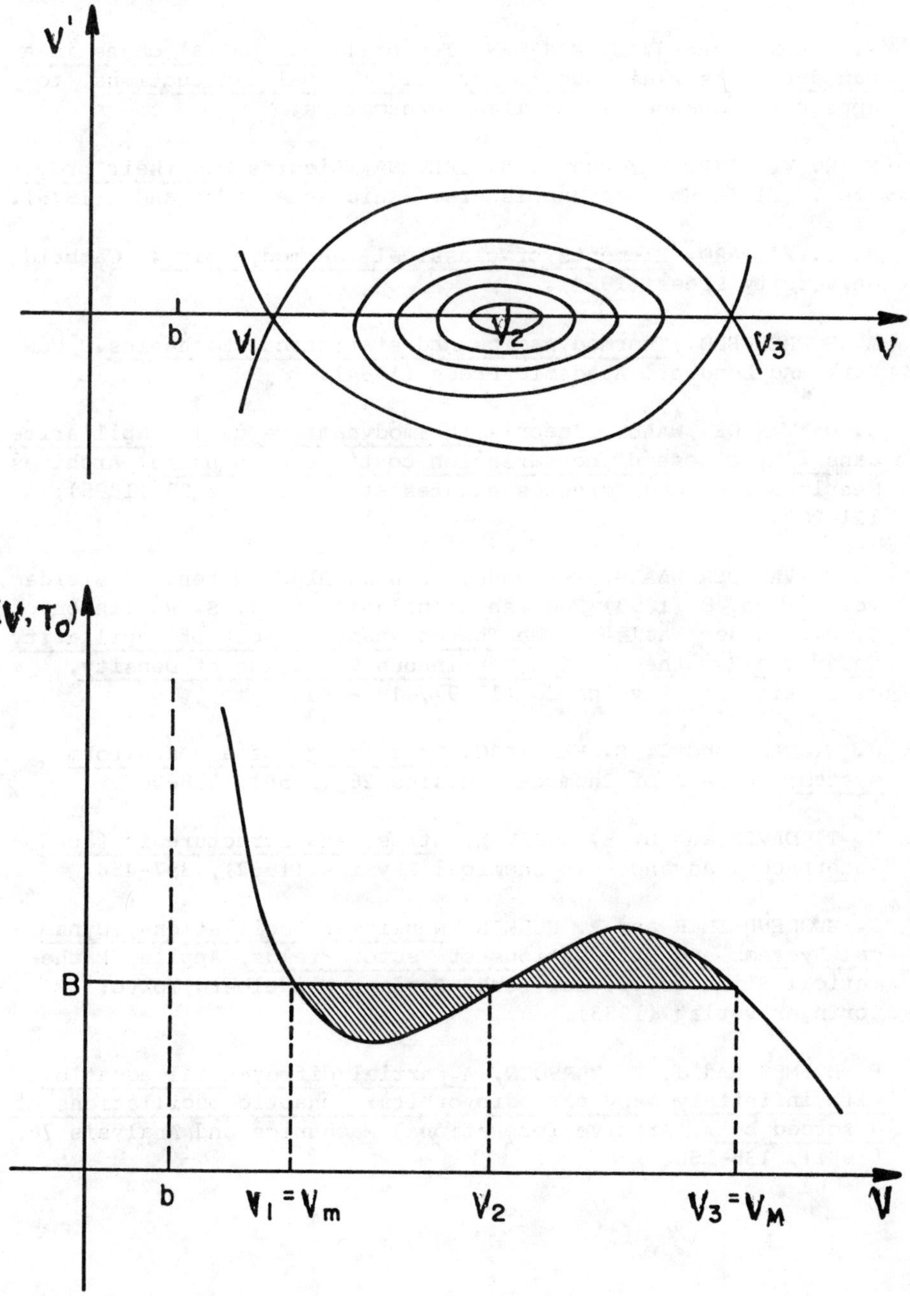

Figure 7.

REFERENCES

[1] M. SLEMROD and J. E. MARSDEN, Temporal and spatial chaos in a van der Waals fluid due to periodic thermal fluctuations, to appear in Advances in Applied Mathematics.

[2] H. N. V. TEMPERLEY and D. H. TREVENA, Liquids and their properties. Ellis Horwood Publishers: Chichester, England, (1978).

[3] A. B. PIPPARD, Elements of classical thermodynamics. Cambridge University Press (1964).

[4] A. SOMMERFELD, Thermodynamics and statistical mechanics. New York and London: Academic Press (1964).

[5] J. D. VAN DER WAALS, Theorie thermodynamique de la capillarite dans l'hypothese d'une variation continue de densite. Archives Neerlandaises des Sciences exactes et Naturelles 28 (1895), 121-209.

[6] J. D. VAN DER WAALS, Verhandel. Konik. Akad. Weten. Amsterdam, vol. 1, no. 8 (1893) English translation by J. S. Rowlinson of J. D. van der Waals' The Thermodynamic Theory of Capillarity Under the Hypothesis of a Continuous Variation of Density, J. Statistical Physics 20 (1979), 197-244.

[7] J. W. CAHN and J. E. HILLIARD, Free energy of a nonuniform system, Journal of Chemical Physics 28 (1958), 258-267.

[8] H. T. DAVIS and L. E. SCRIVEN, Stress and structure in fluid interfaces, Advances in Chemical Physics (1982), 357-454.

[9] J. GUCKENHEIMER and P. HOLMES, Nonlinear Oscillations, Dynamical Systems and Bifurcations of Vector Fields, Applied Mathematical Sciences 42, New York, Berlin, Heidelberg Tokyo: Springer-Verlag (1983).

[10] P. HOLMES and J. E. MARSDEN, A partial differential equation with infinitely many periodic orbits: chaotic oscillations of a forced beam, Archive for Rational Mechanics and Analysis 76, (1981), 135-166.

THE COMPLETE DYNAMICS OF THE FORCED JOSEPHSON JUNCTION CIRCUIT: THE REGIONS OF CHAOS

FATHI M. A. SALAM* AND S. SHANKAR SASTRY**

Abstract. We present the complete dynamics of the Josephson junction circuit with emphasis on the a.c. case. Specifically, we derive analytically the complete bifurcation diagram of the a.c. forced Josephson junction. We thus place on analytic grounds the qualitative, experimental and simulation work of Belykh, Pedersen and Sorensen; specially that which pertains to the regions of chaos. Combining previous results from the literature with our new results, we provide a comprehensive picture of the total dynamics of the a.c. forced case; as well as smooth insightful transitions to the associated I-V characteristics. Explicit asymptotic formulae for the curves that separate the different regions in the bifurcation diagram are also given.

The following is a summary of results in [1] which is to appear in the IEEE Transactions on Circuits and Systems.

1. Introduction. The key to obtaining the a.c. bifurcation diagram is to prove analytically the existence of chaos, for certain parameter values, in the dynamics of the a.c. forced junction. Chaos, i.e. complex orbital behavior, occurs in many systems of practical interest. In addition to a huge volume of stimulating evidence, chaos has been analytically shown to exist, for instance, in the Duffing equations (Guckenheimer and Holmes [7], the swing equations of a power system (Kopell and Washburn [9], Salam, Marsden and Varaiya [3]). It would appear superficially that the results presented in [3] (evidence of the Arnold diffusion variety of chaos) which are valid for system equations associated with the dynamics of forced pendulums, can readily be transcribed to the present case. Certainly the dynamics of the forced junction are those of a forced pendulum. The difference, however, lies in the fact that the damping associated with these dynamics (their departure from being Hamiltonian) is not necessarily small--this necessitates several non-trivial modifications in the theory presented in [7,2,3]. We prove the existence of transversal homoclinic

*Department of Mechanical Engineering and Mechanics, Systems Group, Drexel University, Philadelphia, PA 19104.

**Department of Electrical Engineering and Computer Sciences and the Electronics Research Laboratory, University of California, Berkeley, CA 94720

point, and hence 'Smale horseshoe' chaos in the dynamics of the Josephson junction using the method of Melnikov; thereby validating the experimental and simulation results of [5,6,8] in the 10-300 GHz range[1]. We use our results to obtain the complete bifurcation diagram of the a.c. forced junction.

The outline of our paper and our contributions are as follows: In Section 2 we review the model of the Josephson junction dynamics. We briefly review the bifurcation diagram of the d.c. forced junction derived in [4,6] as well as the I-V characteristic of the d.c. forced junction. We then describe the a.c. bifurcation diagram of [6] derived on the basis of simulation, experiment and qualitative arguments.

We point out in Section 2, the need for analytic proofs for the existence of chaos in certain parameter ranges to analytically confirm the conjecture of [6]. To this end, we begin with a brief discussion of chaos and the Melnikov technique for establishing the presence of a Smale-Birkhoff horseshoe in the dynamics of a periodically forced nonlinear system. We do not review the results on the specifics of the chaos associated with the horseshoe here--the reader is referred to the papers of Kopell and Washburn [9], Salam, Marsden and Varaiya [2], and the new book of Guckenheimer and Holmes [7] for this. In Sections 4 and 5 we apply these techniques to establish the existence of chaos in the junction for different sets of parameter ranges--Section 4 deals with junctions with low conductance values; whereas Section 5 has no such restriction.

We point out here that the treatment of chaos, as it applies to the Josephson junction, was attempted in [11]. The case of large conductance, corresponding to Section 5 here, was only treated qualitatively as in [6]. The attempt to use the Melnikov technique for the low conductance case (with also a zero constant current), corresponding to a special case of Section 4 here, was technically incorrect, for failure to show the following: the improper (Melnikov) integrals exist and are finite; and further they do not equal zero simultaneously. Only in the event that these integrals can be evaluated explicitly, that these technical questions can be directly checked. These points will be clarified in the context of Sections 4 and 5.

In Section 6, we collect all the results to analytically derive the complete a.c. bifurcation diagram of the junction and the relation between the a.c. and d.c. bifurcation diagram. We close the section with a discussion of the effect on our analysis of increasing the amplitude of the a.c. forcing of the junction beyond its 'very small' value.

2. Dynamics of the Josephson Junction

2.1. The Model. The dynamics of a Josephson junction driven by a current source as shown in Fig. 1 (see e.g. [4,5,6,8,10,12]) satisfy the following differential equation:

[1]There may well be other mechanisms, such as period doubling, strange attactors, etc., that generate chaos for the same, or different, parameter ranges from those obtained here.

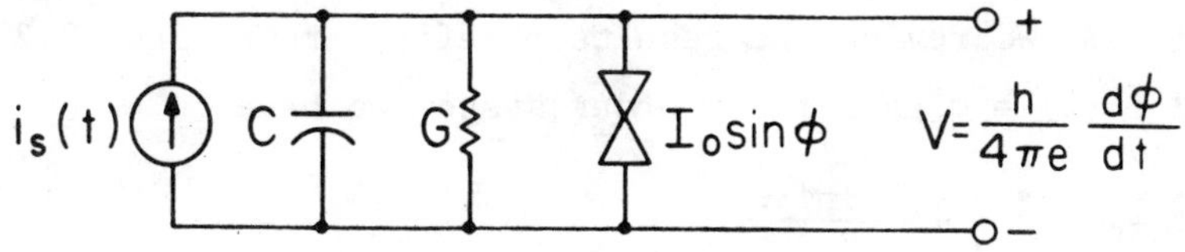

Fig. 1. Josephson junction circuit model.

(2.1) $$\frac{hC}{4\pi e}\frac{d^2\phi}{dt^2} + \frac{hG}{4\pi e}\frac{d\phi}{dt} + I_0 \text{ Sin}\phi = i_s(t)$$

Here h is Planck's constant, e the electronic charge, I_0 a threshold current associated with the tunnelling current, C the junction capacitance, G the junction conductance and ϕ the difference in phase of the order parameters across the junction. The junction voltage v is given by

(2.2) $$v = \frac{h}{4\pi e}\frac{d\phi}{dt} .$$

(2.1) may be rescaled so as to make it dimensionless as follows: set $\tau = \frac{4\pi e I_0}{hG} t$, $\beta = \frac{4\pi e I_0}{h}\frac{C}{G^2}$ and $\bar{i}_s(\tau) = \frac{1}{I} i_s(\frac{hG}{4\pi e I_0}\tau)$. Then (2.1) reads as

(2.3) $$\beta\ddot{\phi} + \dot{\phi} + \text{Sin }\phi = \bar{i}_s(\tau)$$

with $\dot{\phi} = \frac{d\phi}{d\tau}$, $\ddot{\phi} = \frac{d^2\phi}{d\tau^2}$. The form of the scaling (2.3) is not standard (it is degenerate when $G = 0$). Sometimes an alternate scaling of (2.1) is useful. Define

$$\tau' = (\frac{4\pi e I_0}{hC})^{\frac{1}{2}} t; \quad d = (\frac{h}{4eI_0 C})^{\frac{1}{2}} G; \text{ and } i_s'(\tau') = \frac{1}{I_0} i_s((\frac{hC}{4\pi e I_0})^{\frac{1}{2}} t')$$

to obtain

(2.4) $$\ddot{\phi} + d\dot{\phi} + \text{Sin}\phi = i_s'(\tau')$$

(this scaling is degenerate when $C = 0$). The form (2.4) is useful in some contexts since d has the physical interpretation of damping. Note that β of (2.3) is equal to $1/d^2$ in (2.4). We will use both models (2.3) and (2.4) as convenient.

<u>2.2. Constant Forcing (DC Analysis)</u>. Equation (2.3) has been

studied extensively in the instance that $\bar{i}_s(\tau) = \rho$, (equivalently, $i_s(t) = I_{dc}$). We review the results briefly: rewriting (2.3) with $\dot{\phi} = y$ and $\bar{i}_s(\tau) = \rho$ as a first order system we have

(2.5) $\quad \dot{\phi} = y, \; \dot{y} = \dfrac{-y-\sin\phi+\rho}{\beta}$

Equation (2.5) is periodic in ϕ. Consequently, the state (ϕ,y) can be viewed either as an element of $\mathbb{R} \times \mathbb{R}$ or $S^1 = [0,2\pi]$ with 0, 2π identified. The state space $S^1 \times \mathbb{R}$ is more natural, but we use both $\mathbb{R} \times \mathbb{R}$ and $S^1 \times \mathbb{R}$ as per convenience.

2.2.1. Bifurcation Diagram and the IV Characteristics for the DC Excited Junction. The bifurcation diagram of (2.5) with ρ,β as parameters is shown in Figure 2 (for $\beta > 0$). The diagram is symmetric about the β-axis so that we will consider it only for positive values of ρ. In the region (a), equation (2.5) has two equilibrium points in $S^1 \times \mathbb{R}$; the one a node and the other a saddle. All trajectories converge to one or the other equilibrium point. In the region (c), (2.5) has two equilibrium points and a stable periodic orbit on $S^1 \times \mathbb{R}$. The stable periodic orbit on $S^1 \times \mathbb{R}$ corresponds to an unbounded trajectory on $\mathbb{R} \times \mathbb{R}$ - such a periodic orbit is known as a rotation. A trajectory that forms a periodic orbit both on $S^1 \times \mathbb{R}$ and $\mathbb{R} \times \mathbb{R}$ is referred to simply as an oscillation to distinguish it from a rotation.

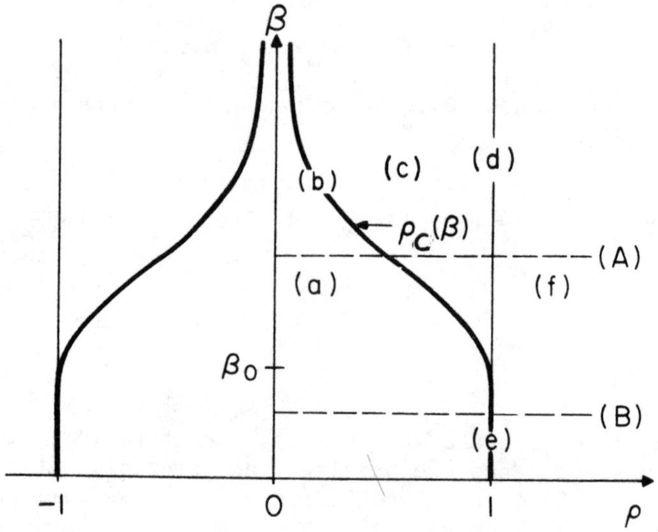

Fig. 2. DC-bifurcation diagram.

In the region (f) (2.5) has only a stable periodic orbit - a rotation. The curves separating regions (a), (c) and (f) are the bifurcation curves - (b): = $\{(\rho,\beta): \rho = \rho_c(\beta), \beta > \beta_o\}$, (d) = $\{(\rho,\beta): \rho = 1, \beta > \beta_o\}$ and (e) = $\{(\rho,\beta): \rho = 1, \beta > \beta_o$. On (b), the phase portrait of (2.5) includes two equilibrium points and a <u>saddle connection</u>. On (d), the phase portrait includes a single equilibrium point obtained by the fusion of the saddle and the node (a <u>saddle-node</u> bifurcation) and a rotation. On the surface (e), the saddle node and saddle connection occur simultaneously. Note that the curves (b), (d) and (e) join smoothly at β_o. The phase portrait $\rho = \beta = 0$ is shown in Figure 4.

While the bifurcation diagram of Figure 2 gives the complete portrait of the dynamics, one is often interested for applications in an I-V characteristic, i.e., a plot of I_{dc} (proportional to ρ, specifically $I_{dc} = I_o$) vs. V_{av} (proportional to the time averaged value of $y = \frac{d\phi}{dc}$, specifically $\frac{h}{2\pi e} < \frac{d\phi}{d\tau} >$). Figure 3 gives two such plots: 3(a) and (b).

Figure 3(a) is obtained by traversing line (A) in the DC-bifurcation diagram (of Figure 2) in the **increasing** ρ direction, starting from $\rho = o$; while Figure 3(b) is obtained traversing line (B). One notes the presence of hysteresis in Figure 3(a) but not in Figure 3(b) - for details see [1].

<u>2.3. Sinusoidal Forcing (AC Analysis)</u>. This is the case of primary interest to us here, namely, equation (2.3) with $\bar{i}_s(\tau) = \rho + \varepsilon \operatorname{Sin}\omega\tau$ (bias + small sinusoidal forcing). It is the model for the dynamics of the Josephson junction when used in microwave generators and mixers [6,8,12]. In standard first order, (2.3) now reads

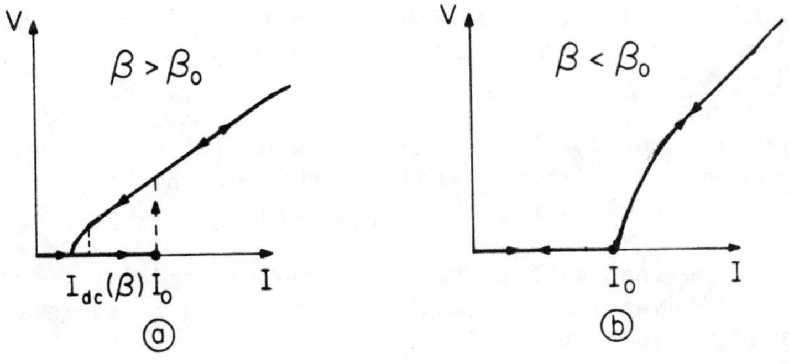

Fig. 3. I-V Characteristic of the DC excited junction.

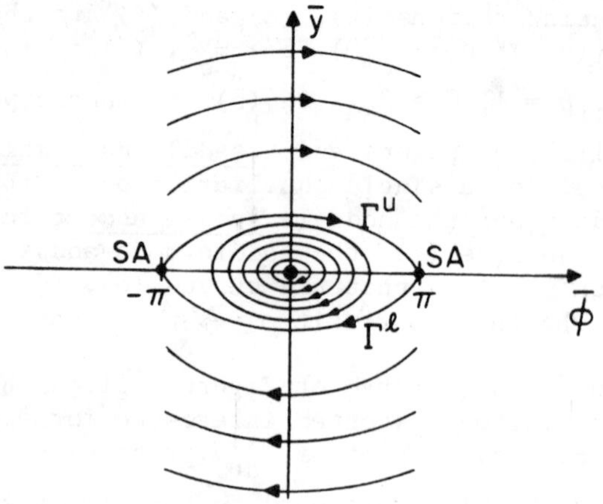

Fig. 4. Phase portraits

$$(2.6) \qquad \dot{\phi} = y, \; \dot{y} = \frac{-y - \mathrm{Sin}\phi + \rho + \varepsilon \; \mathrm{Sin}\omega\tau}{\beta}$$

(2.6) was first studied by Belykh, Pedersen and Sorensen [6]. Their work was thorough but primarily qualitative (i.e., analytic details are often absent). It treated a slightly more general form of equation (2.6), namely,

$$(2.7) \qquad \dot{\phi} = y, \; \dot{y} = \frac{1}{\beta}[\rho - (1+\gamma \cos\phi)y - \mathrm{Sin}\phi + \varepsilon \; \mathrm{Sin}\omega\tau]$$

An AC bifurcation diagram for equation (2.7) was produced in [6] for the parameters β and ρ, and for a fixed (nonzero) γ. This bifurcation diagram contains new regions that exhibit chaotic behavior of trajectories. These regions of chaos were justified via a combination of qualitative arguments and simulation of trajectories. It is needless to say that the delicate question of chaos has to be verified to exist in specific systems such as the one at hand - this point is extremely essential.

We prove rigorously those conclusions of [6] that have not been proven; namely, those that pertain to the regions of chaos. The model we consider is (2.7) with $\gamma = o$ (equivalently, (2.6)).

We emphasize that with these proofs an essentially complete analytic picture of the dynamics emerges. We start with a brief statement of our major technique: the Melnikov Method.

3. Chaos and the Melnikov Method. Consider the system

$$(3.1) \qquad \dot{x} = f(x) + \varepsilon g(x,t)$$

where f, g are sufficiently smooth functions; f from R^2 to R^2; and g from R^3 to R^2 is T-periodic in t. The associated unperturbed system is

$$\dot{\bar{x}} = f(\bar{x}). \tag{3.2}$$

Assume that the system (3.2) possesses a homoclinic orbit $\bar{x}_0(t)$, i.e., an orbit that connects a saddle equilibrium point x_0 to itself. From [7] it can be shown that for ε small enough, the saddle equilibrium point x_0 gets perturbed to a saddle fixed point x_ε of the time-T Poincare map of (3.1).

Under the 'exponential convergence' condition (see [1], [2] for details) the Melnikov integral can be written as an integral of f and g evaluated along the homoclinic orbit of the unperturbed system.

$$M(t_0) = \int_{-\infty}^{\infty} f(\bar{x}_0(t) \wedge g(\bar{x}_0(t), t+t_0). \exp[-\int_{t_0}^{t} \text{trace } D_{f_{\bar{x}}}[\bar{x}_0(\tau)]d\tau]dt. \tag{3.3}$$

Note that the exponential convergence condition is significant only when trace $Df_{\bar{x}}(\bar{x}_0(\tau))]$ is nonzero. If the function $M(t_0)$ has transversal zeros, i.e., if $\exists\ \tilde{t}_0$ such that $M(\tilde{t}_0) = 0$ and $\frac{dM}{dt_0}(\tilde{t}_0) \neq 0$, then it follows that a Smale-Birkhoff horseshoe type of chaos arises in the perturbed system (3.1).

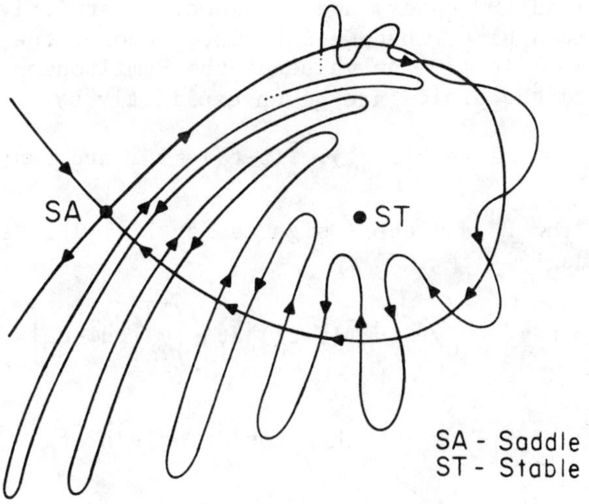

Fig. 5

When the unperturbed system is Hamiltonian, trace $D_{f_{\bar{x}}}(\bar{x}_0(s)) \equiv 0$ so that the Melnikov integral of (3.3) is valid without any further conditions; furthermore the Melnikov integral reads

$$(3.4) \qquad M(t_0) = \int_{-\infty}^{\infty} f(\bar{x}_0(t)) \wedge g(\bar{x}_0(t), t+t_0)dt$$

4. Chaos in the Josephson Junction Dynamics: The Close Hamiltonian Case. We use scaling of equation (2.6), viz.

$$(4.1) \qquad \ddot{\phi} + d\frac{d\phi}{dt} + \sin\phi = i_s'(t),$$

$i_s'(t)$ is of the form $\varepsilon(\rho + \sin\bar{\omega}t)$ and $d = \varepsilon d_0$, where ε is a small parameter. The parameter ranges studied in this section correspond to the large β, small ρ part of the bifurcation diagram of Figure 7. The unperturbed system (for $\varepsilon = 0$) is

$$(4.2) \qquad \begin{aligned} \dot{\bar{\phi}} &= \bar{y} \\ \dot{\bar{y}} &= -\sin\bar{\phi} \end{aligned}$$

(Notation consistent with Section 3). Equations (4.2) are Hamiltonian with (energy) given by

$$(4.3) \qquad H(\bar{y},\bar{\phi}) = \left(\frac{\bar{y}^2}{2} - \cos\bar{\phi}\right).$$

There are two saddle connection orbits for this system as shown in Figure 4 labelled Γ^u (upper) and Γ^ℓ (lower): strictly Γ^u and Γ^ℓ are both homoclinic orbits, when we take into account the 2π - periodicity of equation (4.2) in $\bar{\phi}$. The value of the Hamiltonian $H(\bar{y},\bar{\phi})$ on these orbits is 1 and the orbits are given explicitly by

$$(4.4) \qquad \bar{y}(t-t_0) = \pm 2\,\text{sech}(t-t_0); \quad \bar{\phi}(t-t_0) = \pm 2\arctan[\sinh(t-t_0)].$$

The + sign is for Γ^u and the - sign is for Γ^ℓ. The Melnikov integral (3.4) now reads

$$(4.5) \qquad \begin{aligned} M(t_0) = \int_{-\infty}^{\infty} \bar{y}(\rho - d_0\bar{y} + \sin\bar{\omega}(t+t_0))dt &= \int_{\pm\pi}^{\pm\pi} \rho\, d\bar{\phi} - d_0\int_{-\infty}^{\infty} [\pm 2\,\text{sech}\, t]^2 dt \\ &+ \left[\int_{-\infty}^{\infty} \pm 2\,\text{sech}\, t \cdot \cos\bar{\omega}t\, dt\right]\sin\bar{\omega}t_0 \end{aligned}$$

(We have used here $\frac{d\bar{\phi}}{dt} = \bar{y}$, and the fact that $\int_{-\infty}^{\infty} \text{sech}\, t \sin\bar{\omega}t\, dt = 0$ - integral of an odd function.) Evaluating (4.5) explicitly yields

(4.6) $M(t_0) = \pm\, \rho 2\pi - 8d_0 + [\pi\bar{\omega}\ \mathrm{sech}\ \frac{\pi\bar{\omega}}{2}]\cdot \sin \bar{\omega}t_0$

for Γ^u, the upper homoclinic orbit, the separation

(4.7) $M^u(t_0) = 2\pi\rho - 8d_0 + R(\bar{\omega}) \sin \bar{\omega}t_0$

where $R(\bar{\omega}) := \pi\bar{\omega}\ \mathrm{sech}(\frac{\pi\bar{\omega}}{2}) > 0$. For (4.7) to have a zero $\tilde{t}_0$, we see that it is necessary to have

(4.8) $|-2\pi\rho+8d_0| < R(\bar{\omega})$

It is easy to verify that the zero, $\tilde{t}_0$ is transverse when the inequality is strict for all frequencies $\bar{\omega}$.

For $I_{dc}(:= I_0\rho)$ values satisfying (4.8), the upper homoclinic curve Γ^u breaks up as Figure 6 and hence implies the presence of horseshoe chaos. Analogously for Γ^ℓ, the lower homoclinic orbit, the Melnikov function is

(4.9) $M^\ell(t_0) = 2\pi\rho - 8d_0 - R(\bar{\omega}) \sin \bar{\omega}t_0.$

For (4.9) to have transversal zeros, it is necessary (and sufficient) to have

(4.10) $|2\pi\rho+8d_0| < R(\bar{\omega}).$

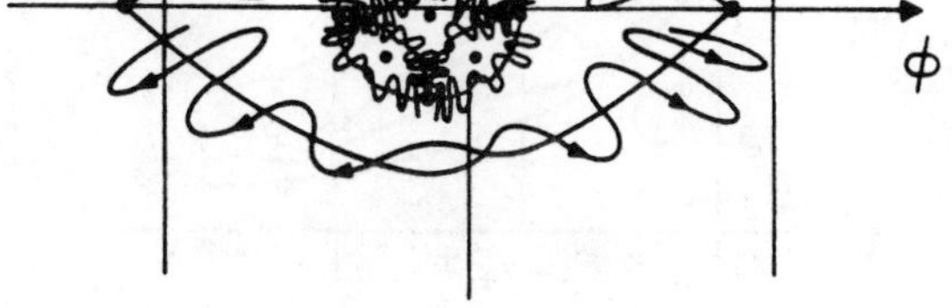

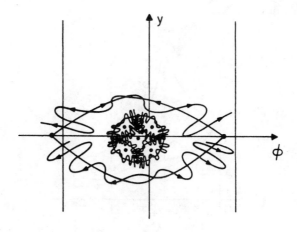

Fig. 6. Small perturbation of the Hamiltonian system.

When ρ is such that both (4.8) and (4.10) are simultaneously satisfied for a fixed ω, we have the complicated phase portrait of Figure 6. (Note that if (4.10) is satisfied, then (4.8) is also satisfied.) In addition to the 'doubly chaotic' intersections of the stable and unstable manifolds of the saddle, there is within the loop of intersecting manifolds an alternating sequence of saddles and stable fixed points of the Poincare map with the manifolds associated with the saddles intersecting each other transversely.

The portrait of Figure 6 persists for a continuum of 'small' ε values corresponding to small d (or to relate this to Figure 7 - large β). The conditions (4.8) and (4.10) enable us to derive the curves ρ_c^-, ρ_c^+ and ρ_0 i.e.,

$$\text{(4.11a)} \qquad \rho > \frac{1}{2}[8\beta^{-\frac{1}{2}}-R(\bar{\omega})] =: \rho_c^-$$

$$\text{(4.11b)} \qquad \rho < \frac{1}{2}[8\beta^{-\frac{1}{2}}+R(\bar{\omega})] =: \rho_c^+$$

Since β is assumed positive, ρ_c^+ is defined for all β, and $\rho_c^+ \to \frac{R(\bar{\omega})}{2\pi}$ as $\beta \to \infty$. Further, at $\beta = (\frac{8}{R(\bar{\omega})})^2$, $\rho_c^- = 0$. The curves ρ_c^- and ρ_c^+ are plotted in Figure 7. Since the analysis of this section is valid only for large and intermediate values of β, the curves ρ_c^-, ρ_c^+ diverge from those predicted by (4.11) (a) and (b) for smaller values of β. Hence in Figure 7, we have shown the actual curves ρ_c^-, ρ_c^+ (the solid lines) diverge from the curves ρ_c^-, ρ_c^+ of (4.11) (a) and (b) (the dotted lines) for small values of β. The discussion of the onset of chaos in the small β case is as in Section 5.

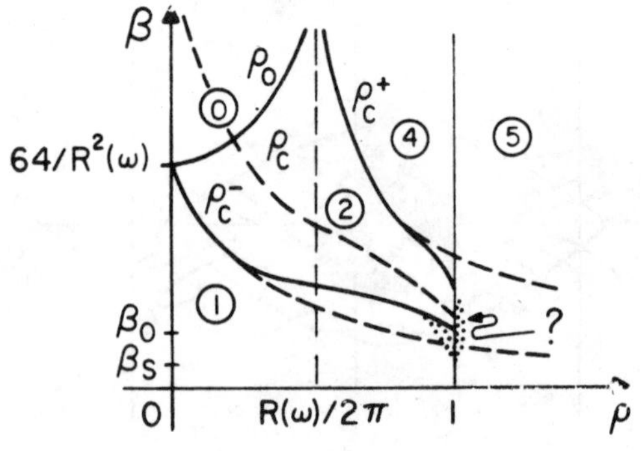

Fig. 7. The AC-bifurcation diagram.

Similarly (4.10) gives the equation for ρ_0, i.e.,

$$(4.12)\qquad \rho > \frac{1}{2\pi}[-8\beta^{-\frac{1}{2}}+R(\bar{\omega})] =: \rho_0$$

Equation (4.12) is defined only for $\beta > \frac{64}{R^2(\bar{\omega})}$ and $\rho_0 \to \frac{R(\bar{\omega})}{2\pi}$ as $\beta \to \infty$. Further at $\beta = (\frac{8}{R(\bar{\omega})})^2$, $\rho_0 = 0$, i.e., it coincides with the curve $\bar{\rho}_c$ at $\beta = (\frac{8}{R(\bar{\omega})})^2$. The curve ρ_0 is plotted in Figure 7 also.

5. <u>Chaos in the Josephson Junction Dynamics: The Non-Hamiltonian Case</u>. The unperturbed system of (2.6), i.e., $\varepsilon = 0$, is

$$(5.1)\qquad \dot{\bar{\phi}} = \bar{y},\ \dot{\bar{y}} = -\frac{\bar{y}-\sin\bar{\phi}+\rho}{\beta} .$$

Let $\rho = \rho_c(\beta)$, i.e., the constant forcing is chosen such that the autonomous system has a homoclinic orbit. The Melnikov integral in this case is given by

$$(5.2)\qquad M(t_0) = \int_{\infty}^{\infty}[\bar{y}(t-t_0)\cdot\frac{1}{\beta}\sin\omega t]\exp(\frac{t}{\beta})dt.$$

(Note that the presence of the term $\exp(\frac{t}{\beta})$ necessitates checking the exponential convergence condition.) This may be simplified as

$$(5.3)\qquad \begin{aligned} M(t_0) = \frac{1}{\beta}\{&[\int_{\infty}^{\infty}\bar{y}(t)e^{t/\beta}\sin\omega t\,dt]\cos\omega t_0 \\ &+[\int_{\infty}^{\infty}\bar{y}(t)e^{t/\beta}\cos\omega t\,dt]\sin\omega t_0\} \end{aligned}$$

If the integrals in the square bracket exist, are finite and are not both zero, then one can easily show that transversal zeros exist (two in each period, $\frac{2\pi}{\omega}$, in t_0) for all but a discrete set of values in ω (i.e., frequencies at which the inequality becomes an equality), see, e.g., Kopell and Washburn [9], Salam, <u>et al</u>. [3]. Since $\bar{y}(t)$ is a bounded smooth function with $\bar{y}(t) \to 0$ as $t \to \pm\infty$, it is enough to show that $\bar{y}(t)e^{t/\beta} \to 0$ fast enough as $t \to \pm\infty$ for the integrals to be finite. Now $\bar{y}(t)$ is a component of $x_0(t)$ the homoclinic orbit of the unperturbed system. Hence for t values close to $\pm\infty$, the rate of approach of $\bar{y}(t)$ to the saddle x_0 is given by the eigenvalues of the linearization of the vector field of the unperturbed system close to the saddle equilibrium $x_0 = (\bar{\phi}_0, 0)$, i.e., the eigenvalues of

$$(5.4)\quad \begin{bmatrix} 0 & 1 \\ -\frac{1}{\beta}\cos\bar{\phi}_0 & -\frac{1}{\beta} \end{bmatrix}$$

where $\bar{\phi}_0$ satisfies $\rho = \sin\bar{\phi}_0$. The eigenvalues of (5.4) are respectively

$$(5.5)\quad \begin{aligned} \lambda^s &= -\frac{1}{2\beta} - \frac{1}{2}[(\frac{1}{\beta})^2 + 4(1-\rho^2)^{\frac{1}{2}}]^{\frac{1}{2}} < -\frac{1}{\beta} < 0 \\ \lambda^u &= -\frac{1}{2\beta} + \frac{1}{2}[(\frac{1}{\beta})^2 + 4(1-\rho^2)^{\frac{1}{2}}]^{\frac{1}{2}} > 0 \end{aligned}$$

where λ^s is the negative (stable) eigenvalue and λ^u the positive (unstable) eigenvalue of the saddle equilibrium point x_0. As $t \to \infty$, $\bar{y}(t)$ approaches 0 as $e^{\lambda^s t}$ while as $t \to -\infty$, $y(t)$ approaches 0 as $e^{\lambda^u t}$. Thus the quantity $\bar{y}(t)e^{t/\beta}$ is of the order of $e^{(\lambda^s+1/\beta)t}$ as $t \to +\infty$ and of the order of $e^{(\lambda^u+1/\beta)t}$ as $t \to -\infty$. From (5.5) it follows that $\bar{y}(t)e^{t/\beta}$ goes to zero exponentially as $t \to \pm\infty$. Hence the integrals are well defined and finite. The Melnikov integral of (5.3) is thus well defined. The only thing that needs to be shown is that the integrals are not both zero at all but discretely many frequencies. This follows from the analyticity of the two integrals in ω.

We have shown the presence of the horseshoe chaos for $\rho = \rho_c(\beta)$. Since the horseshoe is structurally stable, chaos persists for small variations in ρ and β about $\{(\beta,\rho) : \rho = \rho_c(\beta)\ \beta > \beta_0\}$. The curve ρ_c is shown dotted in Figure 7 relative to ρ_c^-, ρ_c^+; and as expected it lies between ρ_c and ρ_c^+. For large values of β the analysis of Section 4 established chaos. For smaller values of β close to β_0; one can only say that a neighborhood of the curve ρ_c exhibits the horseshoe chaos and the boundaries of the chaotic region differ appreciably from ρ_c^- and ρ_c^+ of (4.11a) and (4.11b). We expect that the neighborhood of the curve ρ_c where chaos is encountered shrinks as $\beta \to \beta_0$, since smaller values of β (large vlaues of damping d) increase the rate of convergence of trajectories starting off the saddle connection (of the perturbed system) to the stable equilibrium point (below the saddle connection.)

Acknowledgement. Research sponsored by National Science Foundation Grants ECS-8308330 and ECS-8404723.

REFERENCES

[1] F. M. A. SALAM and S. S. SASTRY, The complete dynamics of the forced Josephson junction circuit: The regions of chaos, Memorandum No. DUMEM SM 83/02, Dept. of Mechanical Engineering and Mechanics, Drexel University, Philadelphia, PA, September 1983.

[2] F. M. A. SALAM, J. MARSDEN and P. VARAIYA, Chaos and Arnold diffusion in dynamical system, IEEE Trans. on Circuits and Systems, CAS-30(1983).

[3] F. M. A. SALAM, J. MARSDEN and P. VARAIYA, Arnold diffusion in the swing equations in a power system, Memorandum No. UCB/ERL M83/13, University of California, Berkeley, March 1983.

[4] A. A. ABIDI and L. O. CHUA, On the dynamics of Josephson junction circuits, Electronic Circuits and Systems, 3(1979), pp. 186-200.

[5] E. BEN-JACOBI, I. GOLDHIRSCH, Y. IRMY and S. FISHMAN, Intermittent chaos in Josephson junctions, Physical Review Letters, 49(1982), pp. 1599-1602.

[6] V. N. BELYKH, N. F. PEDERSEN and O. H. SORENSEN, Shunted Josephson-junction model I - the autonomous case and II. The nonautonomous case, Physical Review B, 16(1977), pp. 4853-4871.

[7] J. GUCKENHEIMER and P. J. HOLMES, Nonlinear Oscillations, Dynamical Systems and Bifurcations of Vector Fields, Applied Mathematical Sciences, No. 42, Springer-Verlag, 1983.

[8] B. A. HUBERMAN, J. P. CRUTCHIELFD and N. H. PACKARD, Noise phenomena in Josephson junctions, Applied Physics Letters, 37(1980), pp. 750-772.

[9] N. KOPELL and R. B. WASHBURN, Chaotic motion in the two-degree-of-freedom swing equations, IEEE Trans. on Circuits and Systems, CAS-29(1982), pp. 738-746.

[10] M. ODYNIEC and L. O. CHUA, Josephson-junction circuit analysis via integral manifolds, IEEE Trans. on Circuits and Systems, CAS-30 (1983).

[11] M. ODYNIEC and L. O. CHUA, Josephson-junction circuit analysis via integral manifolds: Part II, IEEE Trans. on Circuits and Systems, CAS-31 (1984).

BEATING MODES IN THE JOSEPHSON JUNCTION

MARK LEVI*

Abstract

Josephson junction circuits exhibit a rich variety of interesting phenomena. Among such effects are the so-called beating modes (totally different from the conventional beating modes in linear oscillations) in two-point junctions first studied numerically and asymptotically by Y. Imry and L. Schulman. The two-point junction has a transparent mechanical model consisting of two pendula on rods attached to the opposite ends of a horizontal axis which offers an elastic resistance to the twist. The pendula are constrained to the vertical planes perpendicular to the axis, which provides a torsional coupling. The energy is fed into the system by torques applied to the pendula (this corresponds to the current-driven junction), which also encounter a dissipative friction leading to an energy balance. The beating modes correspond to the kinetic energy "jumping" from one pendulum to the other and back - roughly speaking, they "take turns" in rotating.

We give a qualitative analysis of this system, discovering some interesting new phenomena. It turns out, for instance, that the family of systems in question consists of certain equivalence classes, each class indexed by a pair of integers. These integers give a topological characterization of beating modes.

*Boston University, Boston, MA 02215. Supported in part by NSF Grant # MCS-8212681. Part of this work was carried out while the author was a fellow at MSRI at Berkeley.

Introduction.

In this note we describe the dynamics of the system of two coupled pendula

$$(a)\ddot{\phi}_1 + \gamma\dot{\phi}_1 + \sin\phi_1 + k(\phi_1 - \varphi_2) + I$$

$$(b)\ddot{\phi}_2 + \gamma\dot{\phi}_2 + \sin\phi_2 + k(\phi_2 - \phi_1) = 0 \tag{1.1}$$

where $\gamma, k. I$ are constants, in the ranges indicated in section 2 below (e.g., fig. 2.1).

System (1.1) is rich both in its behavior and in its applications, and it has received a large amount of attention, mostly from the physicists. The main stimulus in understanding (1.1) arose from its relevance in studying the Josephson junction, and many experimental and numerical studies [13] [2] have appeared in over two decades since Josephson's discovery [14] in 1962, but to my knowledge, no analytical or geometrical analysis of this basic system has been carried out, with the exception of [2], [4], [9]. Imry and Schulman [2] discovered numerically some remarkable solutions of (1.1), which they called "beating modes";* it is best to describe these in terms of a

Mechanical Interpretation-Coupled Pendula and a Caterpillar

Let two pendula on rods (fig. 1.1) be constrained to rotate in vertical planes perpendicular to a common horizontal rubber axis which provides elastic torsional coupling with elasticity constant k^*: each pendulum is subject to gravity (sin term in (1.1)) and damping with coefficient γ. Moreover, a constant external torque I is applied to one of the pendula. Then the angles $\phi_1 . \phi_2$ formed by the pendula rods with the downward vertical satisfy eq. (1.1).

Now, the above mentioned *beating modes* behave as follows k is small: as one pendulum rests, the other executes several tumbles and then nearly stops, at which point the first pendulum starts and after the same number of tumbles stops; etc. In other words, the pendula "take turns" in rotating. The system absorbs energy from the constant torque in "bursts" (as ϕ_1 tumbles) and spends it in bursts (as ϕ_2 tumbles).

---

* they have nothing to do with the conventional beating modes of the theory of linear oscillations-see Remark 1 below.

* i.e., k is the torque required to twist the axis by 1 rad.

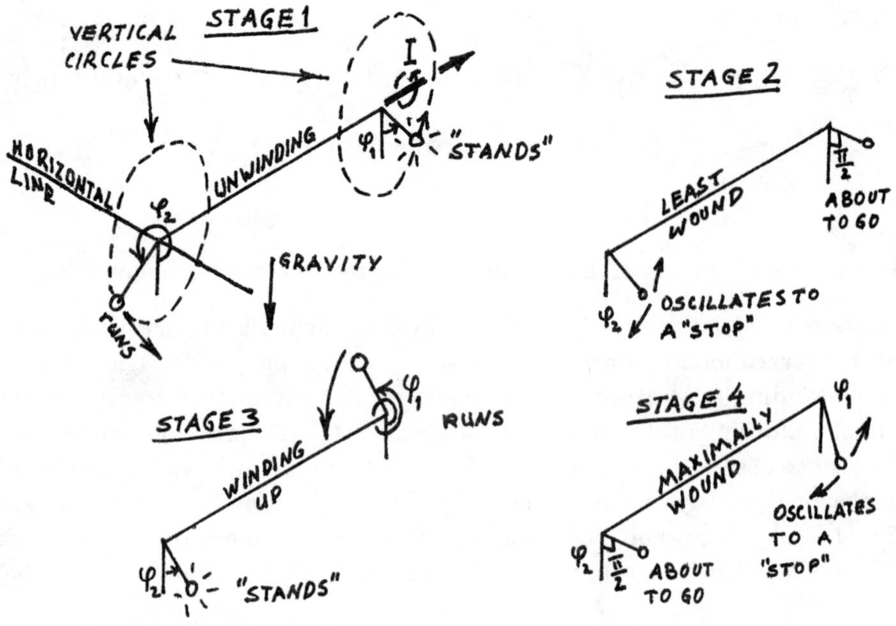

Figure 1.1 Mechanical interpretation of eq. (1.1) and the dynamics of the beating modes.

Caterpillar

Another interesting mechanical analog of (1.1), depicted in fig. 1.2 below, consists of two beads connected by a rubber band of elasticity k.

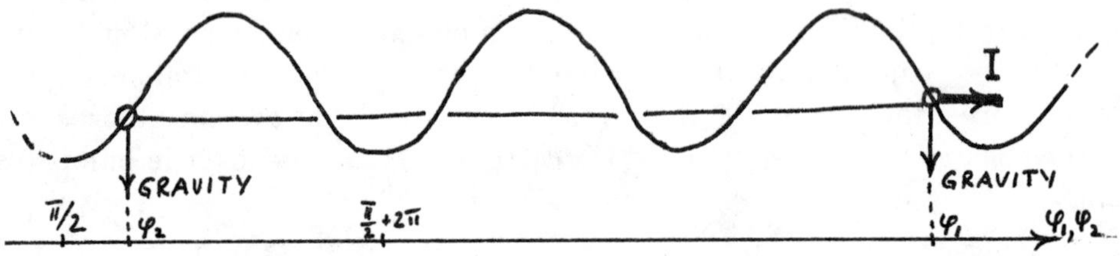

Figure 1.2. Another mechanical interpretation of eq. (1.1).

Each bead slides (with friction) along a certain periodic curve in a vertical plane, and is subject to the friction. The right bead is pulled to the right. Horizontal positions ϕ_1, ϕ_2 of the beads satisfy e.g. (1.1). In terms of this model, beating modes behave like a "caterpillar": as one bead (ϕ_1) moves to the right the distance $0(\frac{1}{k})$, the other nearly rests, and then the roles are reversed, etc.

Interestingly, this behavior occurs with the constant "input" I.

It was observed by Imry and Schulman [2] that beating modes coexist with "synchronous" solutions for which the pendula rotate nearly in unison in contrast to the beating modes.

Remark 1. These "beating modes" are completely different both in shape and (as we shall see) in origin from the standard beating modes in linear oscillations, where one speaks essentially of oscillations near an equilibrium. In our situation the motions are nonlocal-the system is by no means close to an equilibrium and is strongly nonlinear; it is precisely the nonlinearity that gives rise to "beating modes", as we shall explain in section 2.4.

The results mentioned above are numerical and leave open several questions, in particular

(1) what is the mechanism of beating modes?

(2) what is the global behavior of (1.1)?

(3) what are the bifurcations as the parameters change?

The main result of this paper is a qualitative analysis of the dynamics of eq. (1.1) for k small (Theorem 1 in section 2.2).

Our analysis provides some answers to questions (1)-(3), in particular

-a geometrical explanation of the mechanism of the beating modes and of the asymptotic analysis by Imry and Schulman [2].

-a partial description of the geometry of the (four-dimensional) phase space of eq. (1.1) which includes the unstable "invisible" beating mode.

-the bifurcation diagram which breaks up the (I, k)- parameter space into stable mode domains each characterized by a pair of integers with different pairs giving rise to topologically distinct beating modes.

Some questions regarding (2) and (3) are still open and are stated as conjectures in section 2.3. We hope to transform these conjectures into theorems in a forthcoming paper.

We conclude this introductory section with the description of two other situations giving rise to eq. (1.1): one is the Josephson junction and the other-discretization of the sine-Gordon equation.

Josephson Junction

Eq. (1.1) describes the so-called two-point Josephson junction in the current-driven case. This junction is a device consisting of two superconductors separated by two thin gaps. If ϕ_1 and ϕ_2 are

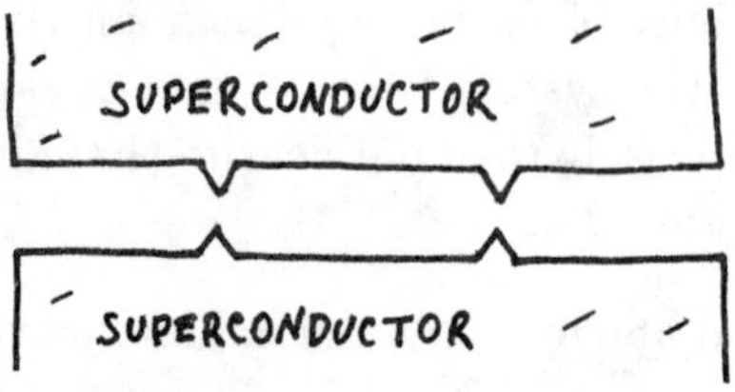

Figure 1.3. A two-point Josephson junction

the jumps in the electron wave function, and if the tunnelling current I is driven across the device, then (in the absence of external magnetic field) ϕ_1 and ϕ_2 obey e.q. (1.1). The voltage across the gap is given by the average velocity of ϕ_1 (or ϕ_2, which is the same). For instance, the equilibrium solutions of (1.1) correspond to the superconducting regime (voltage $= \dot{\phi}_1 = 0$, current $I \neq 0$), while the dynamic solutions ($\dot{\phi}_i \neq 0$) correspond to resistance and hence the energy dissipation.

Further details on the Josephson junction can be found in [3],[7].

Discretization of the sine-Gordon Equation

Eq. (1.1) can be viewed as a (very) discretized version of sine-Gordon equation with damping:

$$\phi_{tt} + \gamma\phi_t - \phi_{xx} + K\sin\phi = 0, \ \ 0 \leq x \leq 1 \tag{1.2}$$

with Neumann boundary conditions

$$\phi_x(0,t) = 0, \ \ \phi_x(1,t) = T, \tag{1.3}$$

cf. [4].

For an integer n, let $h = \frac{1}{n}$ and replace $\phi((i-\frac{1}{2}h, t))$ by $\phi_i(t), i = 1, \ldots, n$. Introducing $\phi_0(t)$ and $\phi_{n+1}(t)$ which would correspond to $\phi(-\frac{h}{2}, t)$ and $\phi(1-\frac{h}{2}, t)$, respectively, we replace the x-derivative in (1.2) and (1.3) by divided differences. Using the discretized boundary conditions to eliminate ϕ_0 and ϕ_{n+1}, we obtain the following system:

$$\ddot{\phi} + \gamma\dot{\phi} + \sin\phi - KA\phi = b, \tag{1.4}$$

where

$$\phi = \text{col}\ (\phi_1, \ldots, \phi_n), \sin\phi = \text{col}(\sin\phi_1, \ldots, \sin\phi_n), b = \frac{1}{h}\ \text{col}\ (0, \ldots, I)$$

and

$$A = \frac{1}{h^2}\begin{pmatrix} 1 & -1 & & & \\ -1 & 2 & -1 & & \\ \vdots & \vdots & & & \\ & & -1\,2 & -1 & \\ & & & -1\,1 \end{pmatrix}$$

For the case $n = 2$ we obtain eq. (1.1). Eq. (1.4) describes a system of n pendula with nearest neighbor interaction.

The question on the qualitative behavior of such higher-dimensional systems will be addressed in a forthcoming paper, with the eventual aim of understanding (as much as possible) the behavior of the damped sine-Gordon equation (1.2)-(1.3).

Acknowledgement. It gives me pleasure to thank Nancy Kopell and Mike St. Vincent for many stimulating conversations in the early stages of this work.

2. Results

In this section we address questions (1)-(3) mentioned in the introduction. In order to state our results precisely (section 2.2), we need some preliminaries (section 2.1), which the reader can skip at first and refer to only when necessary.

2.1 Definitions

Definition 1. We will call a solution (ϕ_1, ϕ_2) of (1.1) a *running periodic solution* (cf. [4]) if there exists $T > 0$ (a *period*) and an integer $m > 0$ such that

$$\begin{aligned} \phi_i(T) &= \phi_i(0) + 2\pi m, \; i = 1,2 \\ \dot{\phi}_i(T) &= \dot{\phi}_i(0) \end{aligned} \tag{2.1}$$

Remark 1. Rewriting (1.1) as a first-order system

$$\begin{aligned} \dot{\phi}_1 &= \psi_1 \\ \dot{\psi}_1 &= -\gamma\psi_1 - \sin\phi_1 - k(\phi_1 - \phi_2) + I \\ \dot{\phi}_2 &= \psi_2 \\ \dot{\psi}_2 &= -\gamma\psi_2 - \sin\phi_2 - k(\phi_2 - \phi_1) \end{aligned} \tag{2.2}$$

in $\mathbf{R}^4 = \{(\phi_1, \psi_1, \phi_2, \psi_2)\}$, we observe that (2.2) is invariant under translation $\tau : (\phi_1, \psi_1, \phi_2, \psi_2) \to (\psi_1 + 2\pi, \psi_1, \phi_2 + 2\pi, \psi_2)$. Identifying the points in $\mathbf{R}^4$ modulo τ turns $\mathbf{R}^4$ into the cylinder $S^1 \times \mathbf{R}^3$; the running periodic solution becomes a periodic one, the integer m from (2.1) giving its homotopy type.

Remark 2. Eq. (1.1) (i.e. (2.2)) has no periodic solutions in $\mathbf{R}^4$ if $\gamma \neq 0$. Equivalently, every periodic solution on the cylinder $S^1 \times \mathbf{R}^3$ has a nonzero homotopy type, i.e. is noncontractable.

Proof. Define the energy of the solution by

$$E(t) = \frac{1}{2}(\dot\phi_1^2 + \dot\phi_2^2) - \cos\phi_1 - \cos\phi_2 - I\phi_1 + \frac{1}{2}k(\phi_1 - \phi_2)^2 .$$

Assuming that a solution (ϕ_1, ϕ_2) of (1.1) is periodic, i.e. that $m = 0$ in (2.1), we run into a contradiction:

$$0 = \int_0^T \dot E(t)dt = \int_0^T [\dot\phi_1(\ddot\phi_1 + \sin\phi_1 + k(\phi_1 - \phi_2) - I) + \dot\phi_2\Big(\ddot\phi_2 + \sin\phi_2 + k(\phi_2 - \phi_1)\Big)]dt =$$

$$-\gamma\int_0^T (\dot\phi_1^2 + \dot\phi_2^2)dt - [\phi_1(T) - \varphi_1(0)] \cdot I < 0,$$

since $\phi_1(T) = \phi_1(0)$.

Definition 2. A running periodic solution (ϕ_1, ϕ_2) of (1.1) is called a *beating mode* if its period consists of two intervals during the first of which ϕ_1 increases by $2\pi m + R_1$, m being an integer and $0 < R_1 < \frac{\pi}{2}$, while ϕ_2 changes by less than $\frac{\pi}{2}$; during the second time interval ϕ_1 and ϕ_2 exchange roles: ϕ_2 increases by $2\pi m + R_2$ and ϕ_1 changes by less than $\frac{\pi}{2}$.

More precisely, $\exists\tau \in (0, T)$ such that

$$\phi_1(\tau) - \phi_1(0) = 2\pi m + R_1,\ 0 < R_1 < \frac{\pi}{2}$$

$$|\phi_2(t) - \phi_2(0)| < \frac{\pi}{2},\quad 0 \le t \le r$$

and

$$|\varphi_1(t) - \varphi_1(\tau)| < \frac{\pi}{2},\quad \tau \le 1 \le T$$

$$\phi_2(T) - \phi_2(\tau) = 2\pi m + R_2,\ 0 < R_2 < \frac{\pi}{2}.$$

Definition 3. A running periodic solution has type (p, q) (p, q are integers, $p > q$) if the distance $\phi_1 - \phi_2$ changes between $2\pi q + R_3$ and $2\pi p + R_4$, $|R_{3,4}| < \frac{\pi}{2}$ during one period, see figure 2.1. In terms of the "caterpillar" model, $2\pi q$ is roughly the shortest length of the crawling caterpillar, and $2\pi p$ is the greatest length.

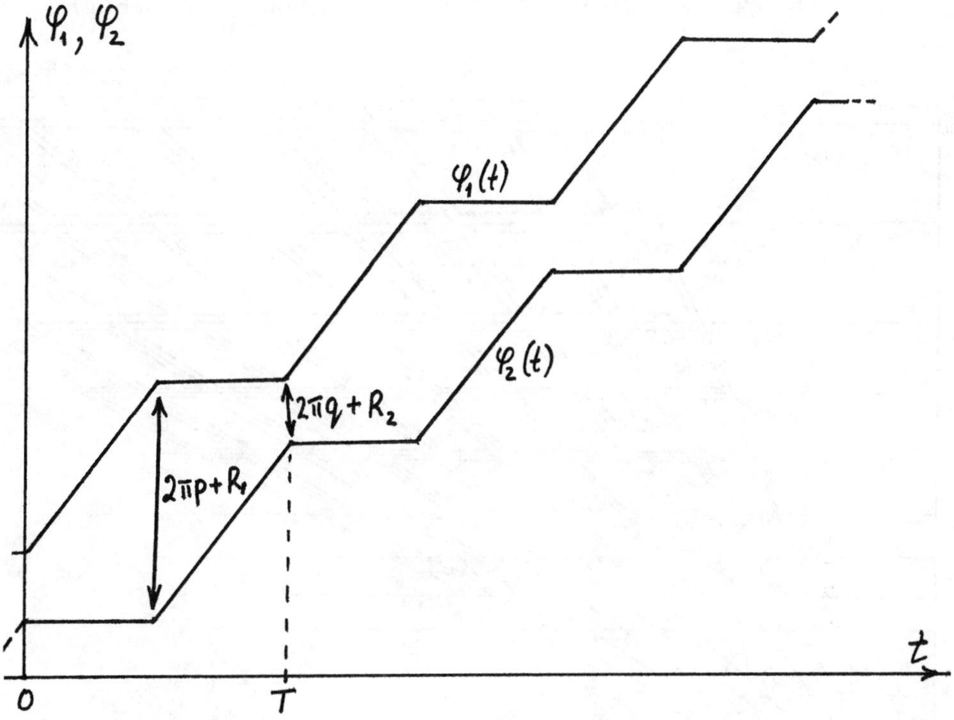

Figure 2.1. A beating mode or "caterpillar" solution.

2.2 Results.

The main result of this paper is the following

Theorem. Fix $\gamma > 0$, and fix two arbitrary constants $C_+ > C_- > 0$ (slopes of the dotted lines in fig. 2.2). There exists $k_o = k_o(\gamma, C_+, C_-)$ such that in the strip $0 < k < k_o$ in the (I,k)-parameter plane there exists a countable sequence of "beating mode domains" $\mathfrak{D}_{p,q}$ enumerated by two integers $p > q$, see fig. 2.2 such that* for any domain $\mathfrak{D}_{p,q}$ which falls into the dotted triangle in fig. 2.2 we have two beating modes of type (p,q).

More precisely, for any $(I,k) \subset \mathfrak{D}_{p,q} \cap$ (dotted triangle) eq. (1.1) possesses *two* beating modes of type (p,q), one exponentially asymptotically stable and the other unstable, of hyperbolic type with a two-dimensional unstable and with a three-dimensional stable manifold.

---

* Precise construction of these domains will follow shortly.

The orbits of these modes in $\mathbf{R}^4$ are within distance π of each other, see fig. 2.3.

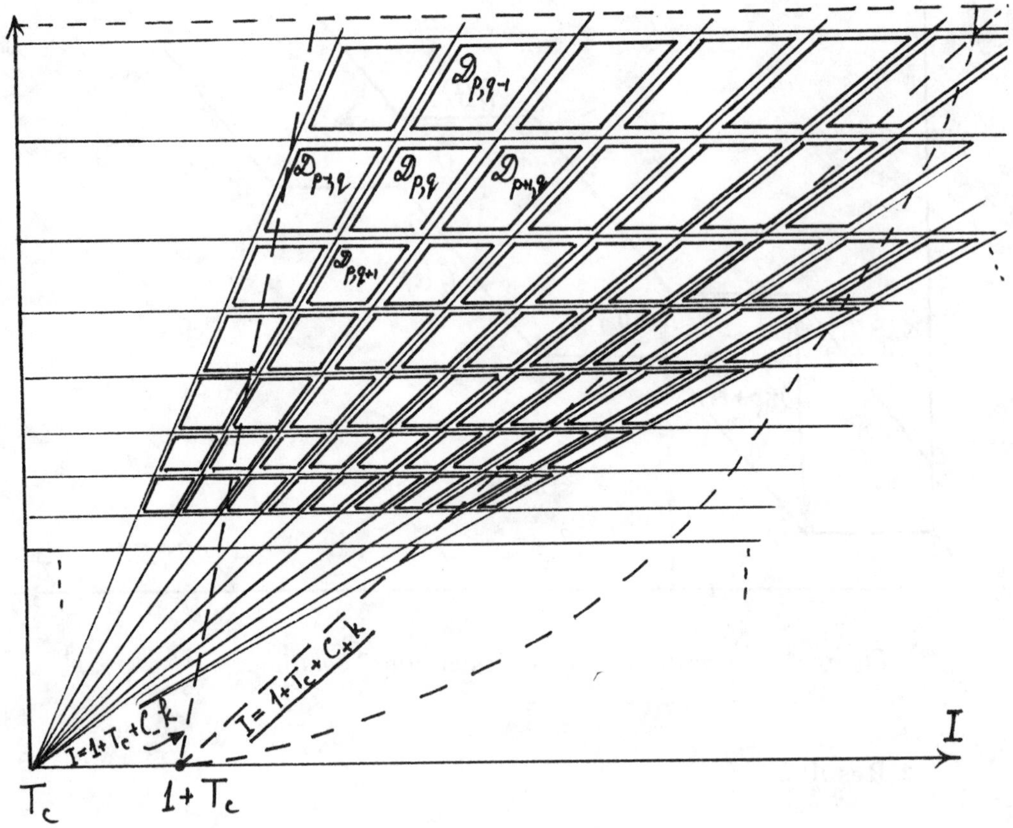

Figure 2.2 Beating mode ("caterpillar") domains.

Before making the structure of the domains $D_{p,q}$ more explicit, we point out a

Remark 3. If a beating mode is of (p,q)-type, then the corresponding orbit winds (p,q) times around the phase cylinder $S^1 \times \mathbf{R}^3$ before closing up. Thus the domains $D_{p,q}$ with different values of the difference $p-q$ give rise to homotopically different beating modes.

Construction of the beating mode domains.

The domains $D_{p,q}$ are defined in terms of the quantities associated with the equation

$$\ddot{\phi} + \gamma\dot{\phi} + \sin\phi + k\phi = a, \quad a = \text{const}, \tag{2.3}$$

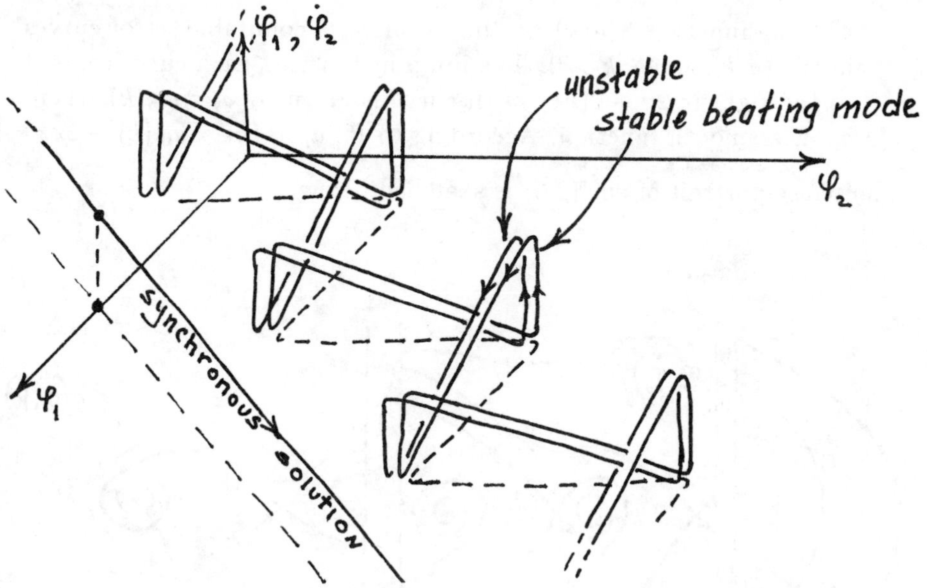

Figure 2.3. Beating modes and the synchronous solution in $\mathbf{R}^4$.

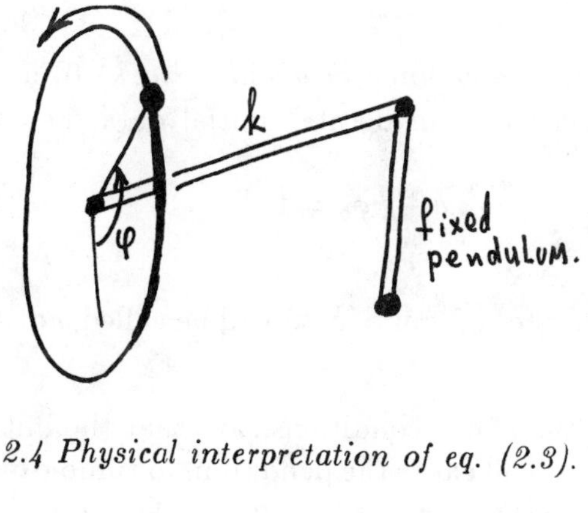

Figure 2.4 Physical interpretation of eq. (2.3).

1°. Periodicity: $S(a + 2\pi k; k) = S(a; k) + 2\pi$.

2°. $\frac{\partial}{\partial a} S(a; k) = \frac{1}{\cos S + k} \geq c(\gamma) > 0$ where $c(\gamma)$ depends only on γ.

3°. $kS(a; k) = a - T_c + \rho(k)$, $|\rho| < 3\pi k$, where $T_c = T_c(\gamma)$ is the critical value of the torque T which gives rise to the saddle-saddle connection for the single pendulum equation $\ddot\phi + \gamma\dot\phi + \sin\phi = T$, see fig. 2.9.

4°. The function $S(a;k)$ has jumps along a countable set of curves in the (a,k)-plane, $k > 0$. The jump in S across each curve crossed to the right is $2\pi + 0(k)$. m-th curve is given by $a = a_m(k)$ where a_m is a smooth function. According to 1°, $a_{m+1}(k) = a_m(k) + 2\pi k$

The phase portrait of eq. (2.3) is sketched below

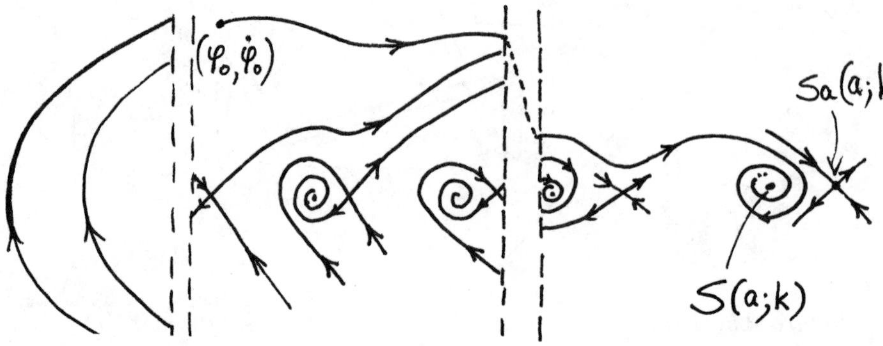

Figure 2.5. Phase Portrait of eq. (2.3).

Definition 4. We define the *distinguished sink* $(S(a;k), 0)$ as that sink of (2.3) which captures the solution with "moderate" initial data, i.e. the data $(\phi_o, \dot{\phi}_o)$ satisfying

$$a - k\phi_o = 1$$

$$\dot{\phi}_o = 1 \tag{2.4}$$

The first saddle point to the right of $S(a;k)$ will be called a distinguished saddle and denoted $Sa(a;k)$.

Intuitively, the first of the above conditions says that the initial net torque on the pendulum equals 1, which will cause the pendulum to tumble many $(0(\frac{1}{k}))$ times, unwinding the axis before settling into a sink. The particular choice of $(\phi_o, \dot{\phi}_o)$ is quite arbitrary; in fact, due to damping the dependence on the initial conditions is very insensitive for most a, k.

To describe the construction of $\mathcal{D}_{p,q}$, we need the

Properties of $S(a;k)$

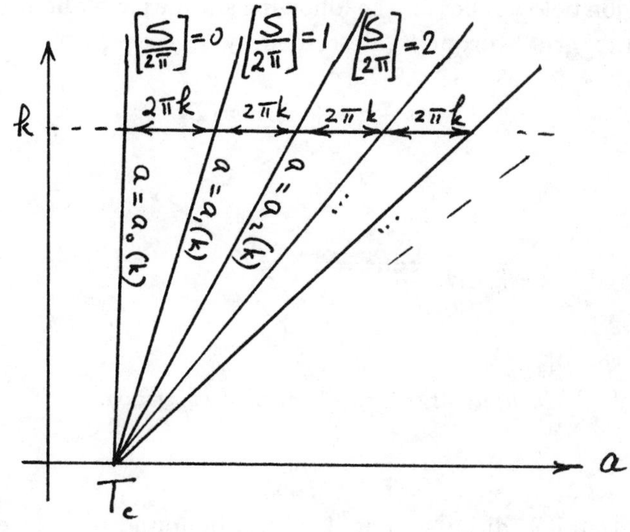

Figure 2.6. Discontinuity lines of $S(a;k)$.

Remark 4. The curves fan out of the point $(T_c, 0)$ as follows from **3°**.

Remark 5. $a_m(k)$ is defined by the requirement that the solution of (2.3) with $a = a_m(k)$ satisfying (2.4) lands on a saddle, fig. 2.5, that is, pt. $(\phi_o, \dot{\phi}_o)$ lies on the stable manifold of a saddle.

Finally, we give the

construction of the domains $\mathcal{D}_{p,q}$. Consider the lines of discontinuity of two functions $S(I + \frac{k\pi}{2}; k)$ and $S(\frac{k\pi}{2}; k)$ in the (I, k)-plane. Each horizontal strip is assigned its own integer $q = -[\frac{1}{2\pi} S(\frac{\pi k}{2}; k)]$; similarly, vertical sectors are enumerated by $p = [\frac{1}{2\pi} S(I + \frac{k\pi}{2}; k)]$, fig. 2.2. The discontinuity lines subdivide the (I, k)-plane $(k > 0)$ into cells; we assign the pair of integers (p, q) to a cell iff this cell is the intersection of p-th sector and q-th horizontal strip. *The beating mode domain* $\mathcal{D}_{p,q}$ is obtained by removing an *αk^2-neighborhood of the boundary of* (p, q)-th cell, where α can be chosen arbitrarily small. To avoid technicalities we can choose $\alpha = \frac{1}{100}$.

Remark 6. Our Theorem holds true even if we delete an exponentially thin neighborhood $e^{-\frac{c}{k}}$ of the boundary of the (p, q)-th cell. We do not give the proof of this statement here to minimize technicalities.

2.3 Heuristic explanation of the results.

A. Single damped pendulum with torque.

Our discussion below relies on the following simple facts about the single damped pendulum with external torque T, described by

$$\ddot{\phi} + \gamma\dot{\phi} + \sin\phi = T, \tag{2.5}$$

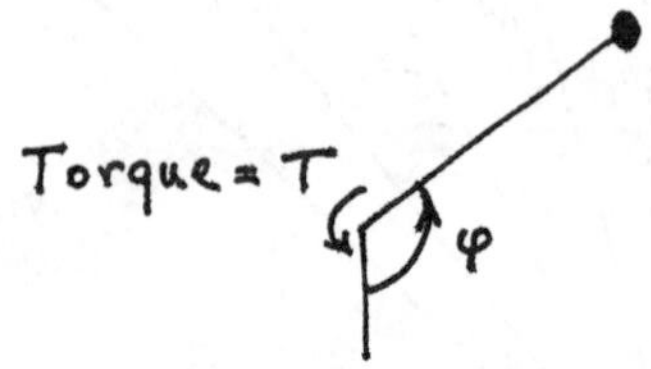

Figure 2.7. Mechanical interpretation of eq. (2.5).

(cf. (2.3)). Detailed discussion of (2.5) can be found in [4]; we only summarize relevant facts. With $\gamma > 0$ fixed there are two bifurcation values for the torque $T : T = 1$ gives rise to the saddle-node bifurcation, fig. 2.8; the other bifurcation

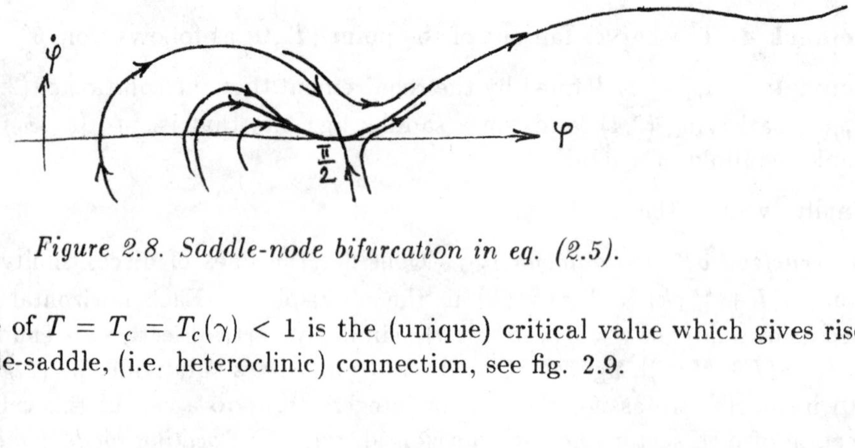

Figure 2.8. Saddle-node bifurcation in eq. (2.5).

value of $T = T_c = T_c(\gamma) < 1$ is the (unique) critical value which gives rise to the saddle-saddle, (i.e. heteroclinic) connection, see fig. 2.9.

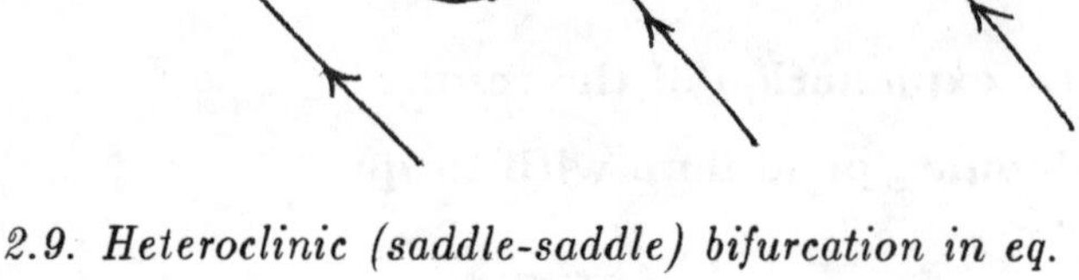

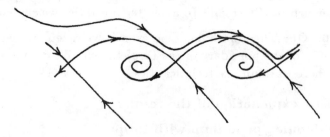

Figure 2.9. Heteroclinic (saddle-saddle) bifurcation in eq. (2.5).

We proceed with the

B. Explanation of the beating modes.

Imry and Schulman [2] have given a partial description of beating modes. There is, however, something unexplained about the phenomenon of one pendulum stopping and the other starting: after all, when one pendulum (say ϕ_1) starts rotating, it twists up the axis thus drawing the other to rotate with it. There is, in other words, an apparent tendency to destroy the alternating character of the beating mode. Numerical and experimental results of Imry and Schulman and Sullivan and Zimmerman [13] suggest that there is some effect that prevents one pendulum from starting too fast thus allowing the other to settle safely into a near-equilibrium.

Here is an informal explanation of this effect of exchange of motion.

Rewriting eq. (1.1) in the form suggested by (2.5):

$$\ddot\phi_1 + \gamma\dot\phi_1 + \sin\phi_1 = I + k(\phi_2 - \phi_1) \quad (a)$$

$$\ddot\phi_2 + \gamma\dot\phi_2 + \sin\phi_2 = k(\phi_1 - \phi_2). \quad (b) \qquad (2.6)$$

we can view (2.6a) and (2.6b) as autonomous systems in the first approximation, since the right-hand sides change slowly: k is small. These equations describe single pendula with slowly varying torques; their phase portraits are shown below.

Say, at $t = 0$ ϕ_2 "stands" and ϕ_1 "runs":

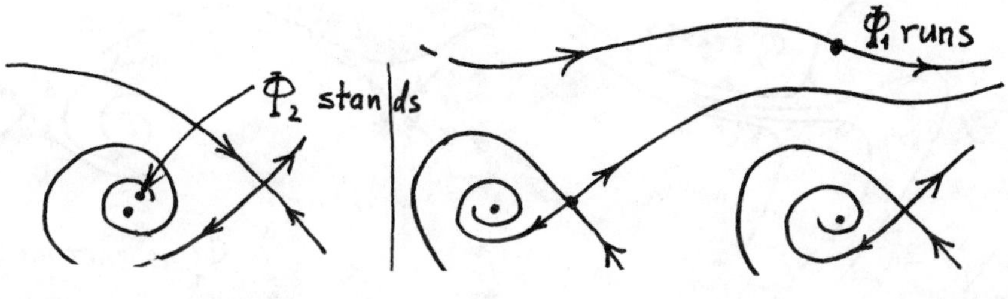

Figure 2.10. Projections of $\Phi(0) = (\phi_1(0), \psi_1(0), \phi_2(0), \psi_2(0))$ onto $\Phi_2 = (\phi_2, \dot\phi_2)$-and (Φ_1)-planes. Phase portraits drawn are those of "frozen" equations: $L\phi_1 = I - k\phi_2(0) \equiv$ const. and $L\phi_2 = k\phi_1(0) \equiv$ const.. Here $L\phi = \ddot\phi + \gamma\dot\phi + \sin\phi + k\phi$.

These two portraits slowly evolve: as ϕ_1 runs, the torque $I - k(\phi_2 - \phi_1)$ (cf. (2.6)) on ϕ_1 decreases, while the torque $k(\phi_1 - \phi_2)$ acting on ϕ_2 grows. Geometrically, this causes the sink and the saddle in fig. 2.10 in the $\Phi_2 = (\phi_2, \dot\phi_2)$-plane to approach each other, while the running periodic solution in the $\Phi_1 = (\phi_1, \dot\phi_1)$-plane moves down. If a beating mode is to exist, the sink and the saddle must coalesce at about the same time as ϕ_1 stops, i.e., when the running periodic solution of eq. (2.6a) touches the saddle points:

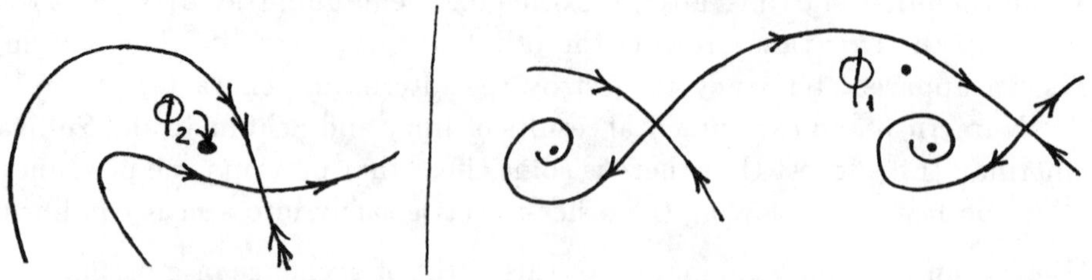

Figure 2.11. In order for the exchange of motion between two pendula to occur the system must not be too far from the simultaneous bifurcation shown here.

Existence of beating modes requires therefore a nearly simultaneous bifurcation to occur: a saddle-node in the Φ_2-plane and a saddle-saddle connection in the Φ_1-plane. More specifically, if a beating mode is to exist, Φ_1 will enter a small neighborhood of the sink and only then will Φ_2 start, fig. 2.12.

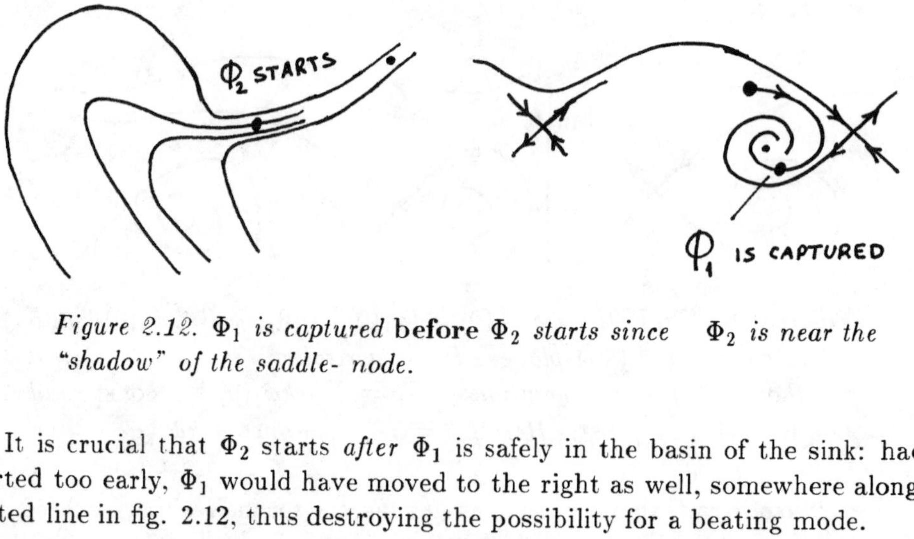

Figure 2.12. Φ_1 is captured **before** *Φ_2 starts since Φ_2 is near the "shadow" of the saddle- node.*

It is crucial that Φ_2 starts *after* Φ_1 is safely in the basin of the sink: had Φ_2 started too early, Φ_1 would have moved to the right as well, somewhere along the dotted line in fig. 2.12, thus destroying the possibility for a beating mode.

In fact, we will show that with $(I, k) \in \mathcal{D}_{p,q}$ the beating mode exists and is such that Φ_2 stays near $(\frac{\pi}{2}, 0)$ long enough $(0(\frac{1}{\sqrt{k}}))$ to allow the time for Φ_1 to settle near a sink-that time is of order $0(ln\frac{1}{k}) << 0(\frac{1}{\sqrt{k}})$.

Summaring, it is the slow starting of a pendulum that is responsible for the existence of beating modes. Now, the reason for this slow starting is simple: Φ_2 starts where the sink and saddle have coalesced, i.e. Φ_2 starts in a shadow of a *degenerate* equilibrium point (saddle-node) and thus takes a long time to depart. On the other hand, for ϕ_1 to pass by the saddle point takes a relatively short time $(0(ln\frac{1}{k}))$ provided Φ_1 is not too close to the stable manifold. The latter condition is assured by the choice of $(I,k), \in \mathfrak{D}_{p,q}$, i.e. by the definition of $\mathfrak{D}_{p,q}$.

This was a rough explanation of the nature of the beating modes.

A more precise discussion and proofs are given in a forthcoming paper. We conclude this subsection with the

C. heuristic explanation of the diagram of beating mode domains $\mathfrak{D}_{p,q}$.

The horizontal lines in fig. 2.2 i.e. the lines of discontinuity of $S(\frac{k\pi}{2};k)$ correspond to those values of k for which the solution of $\ddot{\phi}+\gamma\dot{\phi}+\sin+k\phi=\frac{k\pi}{2}$ with the initial data (2.4) lands on a saddle. That is to say, for those k the solution of the actual equation

(2.1a): $\ddot{\phi}_1+\gamma\dot{\phi}_1+\sin\phi_1+k\phi_1+k\phi_2$, when $\phi_2\approx\frac{\pi}{2}$ will pass too close to the saddle, thus destroying the chances for a beating mode to exist. This explains the reason for deleting neighborhoods of horizontal lines in fig. 2.2; the neighborhoods of the sloped lines are deleted for the same reason. In conclusion we give a crude explanation of the restriction of $\mathfrak{D}_{p,q}$ to the triangle in fig. 2.2.

The existence of beating modes requires two simultaneous near- bifurcations, as we saw above (fig. 2.11):

(1) a saddle node for one pendulum, say ϕ_2:

$k(\phi_1-\phi_2)\approx 1.$

and (2) a heteroclinic (saddle-saddle) connection for the other:

$$I+k(\phi_1-\phi_2)\approx T_c.$$

Note that the above expressions are the net external torques acting on the individual pendula, i.e. the right hand sides of (2.6).

From the above estimates we obtain

$$I\approx 1+T_c.$$

cf. [2]. Actually, the laxity of this approximate requirement depends on k.

The discussion above is very sketchy; precise statements and proofs are given in a forthcoming paper.

2.4 Open Questions

We state several hypothetical properties of te system as conjectures, hoping to transform these into theorems in the future.

Conjecture 1. Three-dimensional stable manifold of the hyperbolic beating mode separates $\mathbf{R}^4$ into two components: the stable beating mode together with all the equilibria* lies in one component, while the synchronous solution lies in the other.

Remark 8. Mike St. Vincent has recently given a complete analysis of the equilibria of (1.1), combining all the information in a single elegant pictue [15].

Conjecture 2. As (I, k) crosses the boundary of domain $\mathcal{D}_{p,q}$ within the triangle in fig. 2.2, the two (p, q) type beating modes coalesce and disappear in a saddle-node bifurcation. As

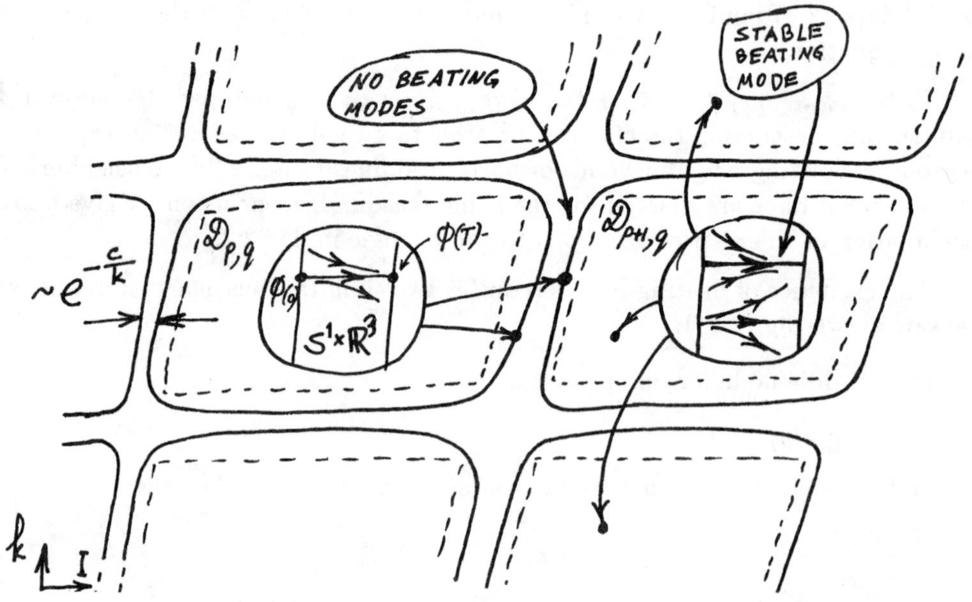

Figure 2.14. Beating modes coalesce in a saddle-node bifurcation and disappear as (I, k) leaves the beating mode domain. As (I, k) crosses "the street" and enters another domain, a new pair of beating modes appears. The modes are sketched very schematically; e.g., the $p - q$ windings around the "hole" in $S^1 \times \mathbf{R}^3$ are not shown.

(I, k) crosses over into a neighboring domain $\mathcal{D}_{p',q'}$, a pair of (p', q')-type beating modes is born as a result of another saddle node bifurcation, fig. 2.14.

* It is not hard to see that eq. (1.1)has $O(k^{-1})$ equilibrium points.

References

[1] M. Cirillo, R.D. Parmentier and B. Savo. Mechanical analog studies of a perturbed sine-Gordon equation. *Physica* **3D** (1981), 565-576.

[2] Y. Imry and L. Schulman, Qualitative theory of the nonlinear behavior of coupled Josephson junctions. *J. Appl. Phys.* **49(2)**, February 1978, 749-758.

[3] D.N. Landberg, D.J. Scalapino and B.N. Taylor, The Josephson effects, *Scientific American*, **21**, 30-39.

[4] M. Levi, F.C. Hoppensteadt and W.L. Miranker, Dynamics of the Josephson Junction *Quarterly of Applied Mathematics*, July 1978, 167-198.

[5] P.M. Marcus and Y. Imry, Steady oscillatory states of a finite Josephson junction, *Solid State Communications*, **33**, 1980, 345-349.

Solid State Communications, **41**, **No.2**, 1982, 161-166.

[7] J. Matisoo, Josephson-type superconductive tunnel junctions and applications, *IEEE Transactions on Magnetics*, **5**, 1969, 848-873.

[8] Nakajima, K., T. Yamashita, and Y. Onodera, Mechanical analogue of active Josephson transmission line, *Journal of Applied Physica*, **45**, **No.7**, July 1974, 3141-3145.

[9] F. Odeh. On existence, uniqueness and stability of solutions of the Josephson phase equation, to appear.

[10] A.C. Scott. A nonlinear Klein-Gordon equation *American Journal of Physics*, **37**, **No.1**, January 1969, 52-61.

[11] M.P. Soerensen, N. Arley, P.L. Christiansen, R.D. Parmentier, and O. Skovgaard, Intermittent switching between soliton dynamic states in a perturbed sine-Gordon model. *Phys. Rev. Letters*, November 1983, 1919-1922.

[12] D.B. Sullivan, J.E. Zimmerman, Mechanical analogs of time dependent Josephson phenomena, *Amer. J. Phys.*, **39**, **No.12** (Dec. 1971), 1504-1517.

[13] F. Tricomi, Integrazione di un'equazione differenzale presentasi in elettrotecnica, *Ann. Sc. Norm. Sup. Pisa* **2** (1933).

[14] J.E. Zimmerman and D.B. Sullivan, High-frequency limitations of the double-junction SQUID amplifier, *Applied Physics Letters*, **31**, **No.5** September 1977, 360-362.

[15] B.D. Josephson, *Phys. Lett.*, **1** (1962), 251.

[16] M. St. Vincent, a private communication.

SOLITONS AND CHAOTIC INTERMITTENCY IN THE SINE-GORDON SYSTEM MODELLING THE JOSEPHSON OSCILLATOR

PETER L. CHRISTIANSEN*

Abstract. Long Josephson tunnel junction oscillators are accurately modelled by the perturbed sine-Gordon system. Soliton dynamic states for this system, hysteresis and chaotic intermittency between the states have been found by computer experiments.

1. Introduction. The perturbed sine-Gordon equation [1] models the dynamics of long Josephson tunnel junctions. Thus the detailed structure of the current-voltage (I-V) characteristics of such junctions, the emission of microwave radiation from junctions and the properties of the radiation seem explainable in terms of the soliton dynamics resulting from the perturbed sine-Gordon equation. The paper discusses some of the soliton dynamic states for the perturbed sine-Gordon system. Our recent computer experiments exhibiting hysteresis phenomena and chaotic intermittency between soliton dynamic states occurring as a result of applied external bias current and magnetic field are also discussed. The perturbed sine-Gordon system is a suitable testing ground for coherence effects and chaotic effects from recent non-linear dynamics. These effects and their balance determine the dynamical properties of the technologically important Josephson oscillator.

2. Modelling the Josephson Oscillator. A Josephson tunnel junction consists of two superconducting metal layers separated by a thin insulating oxide layer of uniform thickness (t_{OX}) that is small enough to permit quantum-mechanical tunnelling of electrons. Figure 1 shows a Josephson tunnel junction oscillator in which the two layers overlap in a region of length L in the X-direction and width W in the Y-direction. When L >> W the junction is essentially one-dimensional in space. The tunnelling supercurrent is

(1) $j(X,T) = j_0 \sin\phi,$

where $\phi = \phi(X,T)$ is the difference between the phases of the order parameter of the two superconductors, j_0 is the maximum current, and

*Laboratory of Applied Mathematical Physics,
The Technical University of Denmark,
DK-2800 Lyngby,
Denmark.

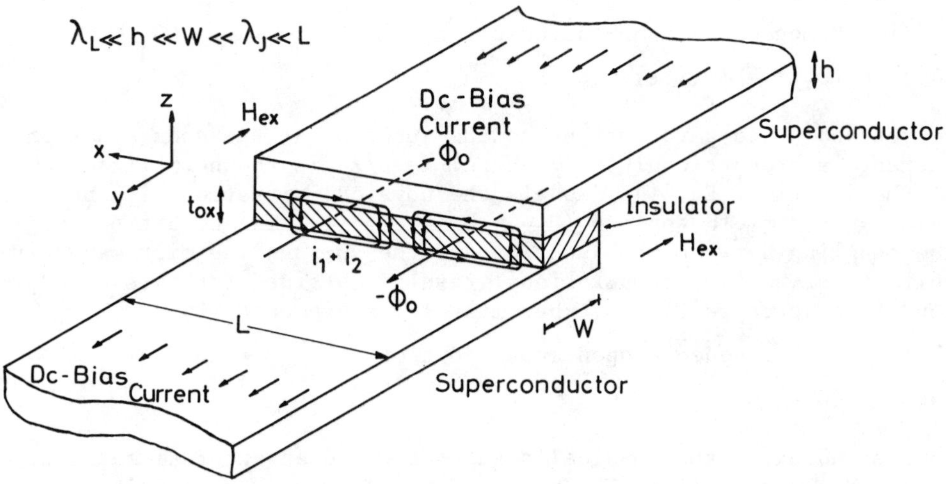

Figure 1. Josephson tunnel junction oscillator of overlap geometry. [2].

T is laboratory time. The voltage drop across the insulating layer is

$$(2)\quad V(X,T) = \frac{\hbar}{2e}\frac{\partial\phi}{\partial T}$$

where $\hbar$ is Planck's constant divided by 2π and e is the electronic charge. Combination of Eqs. (1 - 2) and Maxwell's equations yields the perturbed sine-Gordon equation [2]

$$(3)\quad \phi_{xx} - \phi_{tt} - \sin\phi = \alpha\phi_t - \beta\phi_{xxt} - \gamma$$

in normalized coordinates $x = X/\lambda_J$ and $t = T\omega_0$. The Josephson length is given by $\lambda_J = (\Phi_0/2\pi\, j_0\, Lp)^{1/2}$ and the Josephson plasma frequency $\omega_0 = (2\pi\, j_0/\Phi_0\, C)^{1/2}$. Here $\Phi_0 = h/2e$ is the magnetic flux quantum and Lp and C are inductance and capacitance per length unit of the junction respectively. The perturbation terms, $\alpha\phi_t$ and $-\beta\phi_{xxt}$, represent quasiparticle tunnelling loss and surface impedance loss respectively. Thus the coefficients $\alpha = G/\omega_0 C$ and $\beta = \omega_0\, Lp/Rp$ where G^{-1} and Rp are the normal resistance and the surface resistance per length unit of the junction respectively. In the term $\gamma = j_B/j_0$, j_B is the externally applied bias current per length unit. Typical values of λ_J and ω_0 are $\lambda_J = 1.56\times10^{-4}$ m and $\omega_0 = 5.8\times10^{10}\ s^{-1}$ such that the propagation velocity for electromagnetic signals (i.e. solutions of the linear wave equation, $\phi_{xx} - \phi_{tt} = 0$) becomes $c = \lambda_J\,\omega_0 = 9.05\times10^6$ m/s in laboratory coordinates along the junction. In the normalized coordinates this velocity is of course equal to unity.

At the ends of the junction, $X = 0$ and $X = L$, we apply the following sets of boundary conditions:

(i) Homogeneous open-end conditions

$$(4a)\quad \phi_x(0,t) = \phi_x(\ell,t) = 0$$

corresponding to zero current on the junction at the ends since the current is proportional to ϕ_X. Here $\ell = L/\lambda_J$ is the normalized length of the junction. Eq. (4a) models the physical situation with no external magnetic field applied ($H_{ex} = 0$ in Fig. 1). The conditions neglect the coupling between the junction and the surrounding microwave circuit. Nevertheless, computational results obtained for this boundary condition agree well with experimental measurements [2].

(ii) Inhomogeneous open-end conditions

$$(4b)\quad \phi_x(0,t) = \phi_x(\ell,t) = \eta$$

corresponding to the physical situation where an external magnetic field, H_{ex}, is applied ($H_{ex} \neq 0$ in Fig. 1). Here $\eta = (-w/j_0\ \lambda_J)H_{ex}$.

The initial conditions for the computational modelling of the oscillator are

$$(5)\quad \phi(x,0) = F(x) \quad \text{and} \quad \phi_t(x,0) = G(x)$$

where the functions F and G are chosen such that stationary states of $\phi(x,t)$ are obtained in the numerical computations without too long transients. In practice the final ϕ and ϕ_t distributions from a nearby stationary state in parameter space $(\alpha,\beta,\gamma,\ell,\eta)$ are often used as the functions F and G respectively. The influence of some small perturbations on the stationary state is illustrated in Ref. [3].

3. Soliton Dynamic States. The classical sine-Gordon equation, $\phi_{xx} - \phi_{tt} - \sin\phi = 0$, has 2π-kinks and anti-kinks

$$(6)\quad \phi(x,t) = 4\tan^{-1}[\exp(\pm(x - ut - x_0)/\sqrt{1-u^2}]$$

as soliton solutions [4]. Here u is the constant velocity of the soliton and x_0 is the position of the soliton at $t = 0$.

The perturbed sine-Gordon equation (3) has similar soliton solutions in the looser sense. Each soliton carries a magnetic flux quantum. The dynamics of these solitons is investigated by means of perturbation theory in Ref. [1]. As a result a first-order differential equation for the variable soliton velocity for a single soliton, u(t), is derived

$$(7)\quad \frac{du}{dt} = \pm\frac{1}{4}\pi\gamma(1-u^2)^{3/2} - \alpha u(1-u^2) - \frac{1}{3}\beta u.$$

Eq. (7) expresses the balance between energy input in the system due to the γ-term and dissipation due to the loss terms, $\alpha\phi_t$ and $-\beta\phi_{xxt}$. The stationary velocity, u_∞, is determined from Eq. (7) by letting

$du/dt = 0$ and solving the resulting equation with respect to u. In typical computer experiments u(t) rapidly adjusts towards the stationary velocity, u_∞.

For a finite junction with open-end boundary conditions (4a) it is easy to show that solitons are reflected into antisolitons and vice versa at the boundaries. The bias current, γ, drives the soliton in the negative x-direction until it is reflected into an antisoliton at $x = \ell$. The antisoliton is driven in the positive x-direction and reflected into a soliton at $x = 0$ and a new cycle of this stationary state is initiated. We shall designate such a stationary state a soliton dynamic state. The periodic motion of the soliton on the oscillator is responsible for the emission of electromagnetic radiation, typically in the GHz-range, from the oscillator. Figure 2 shows a com-

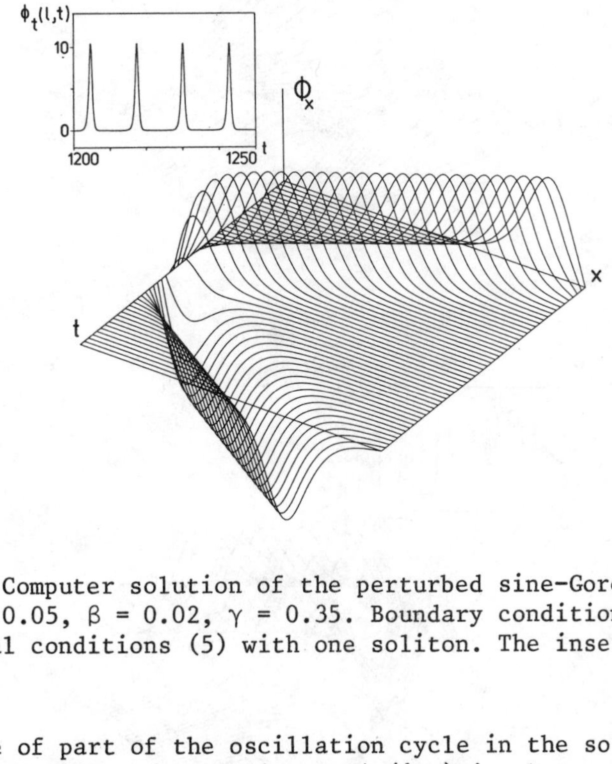

Figure 2. Computer solution of the perturbed sine-Gordon equation (3) with $\alpha = 0.05$, $\beta = 0.02$, $\gamma = 0.35$. Boundary conditions (4a) with $\ell = 6$. Initial conditions (5) with one soliton. The inset shows $\phi_t(\ell,t)$ [2].

puter picture of part of the oscillation cycle in the soliton dynamic state with one soliton. In the inset, $\phi_t(\ell,t)$ is shown as a function of t for 50 time units. This quantity is proportional to the voltage on the oscillator at the end at $x = \ell$ according to (2). The DC-component of the voltage $\phi_t(\ell,t)$ has been computed for different values of the applied bias current γ in Eq. (3). The resulting curve shows agreement with experimentally measured IV-curves for the junction [5]. Each soliton dynamic state corresponds to a branch of the IV-characteristic. These branches are designated zero field steps (ZFS) in the physical literature. Computational Fourier analysis of $\phi_t(\ell,t)$ provides the power spectrum for the radiation from the oscillator. The basic frequency f is given by

(8) $f = u/2\ell$.

Also the computational power spectrum shows agreement with experimental measurements of the power spectrum, [5] and [6].

The presence of the loss term $-\beta\phi_{xxt}$ in Eq. (3) permits soliton dynamic states in which two or more solitons travel together in bunches [1].

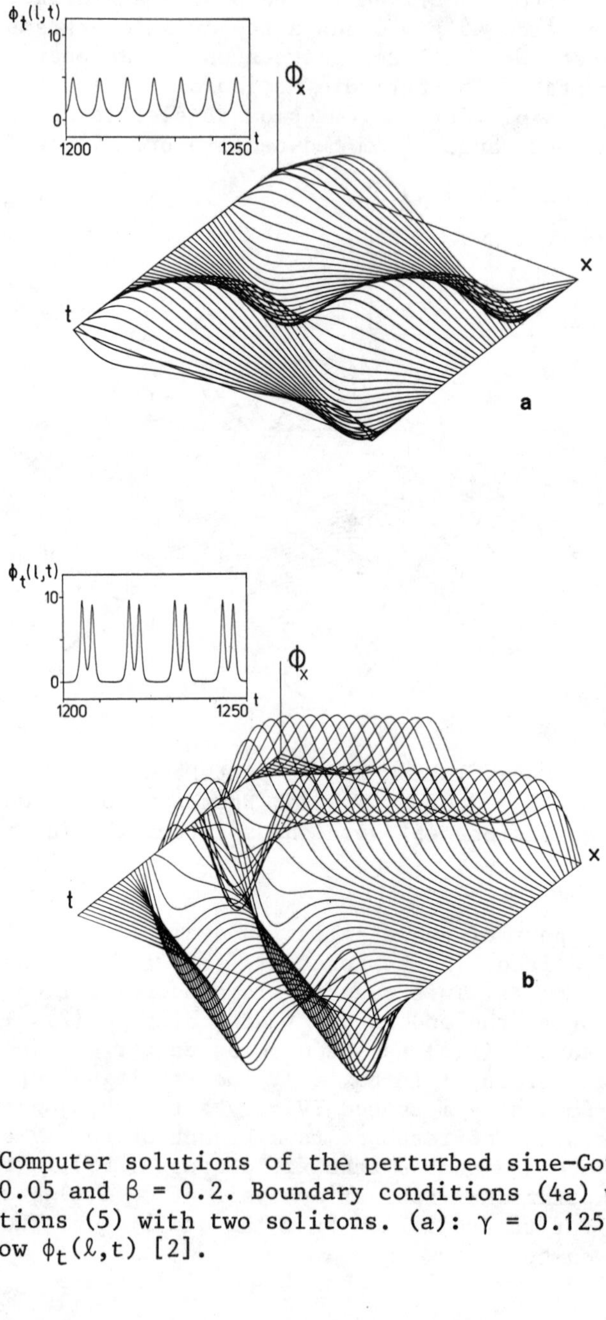

Figure 3. Computer solutions of the perturbed sine-Gordon equation (3) with $\alpha = 0.05$ and $\beta = 0.2$. Boundary conditions (4a) with $\ell = 6$. Initial conditions (5) with two solitons. (a): $\gamma = 0.125$, (b): $\gamma = 0.3$. The insets show $\phi_t(\ell,t)$ [2].

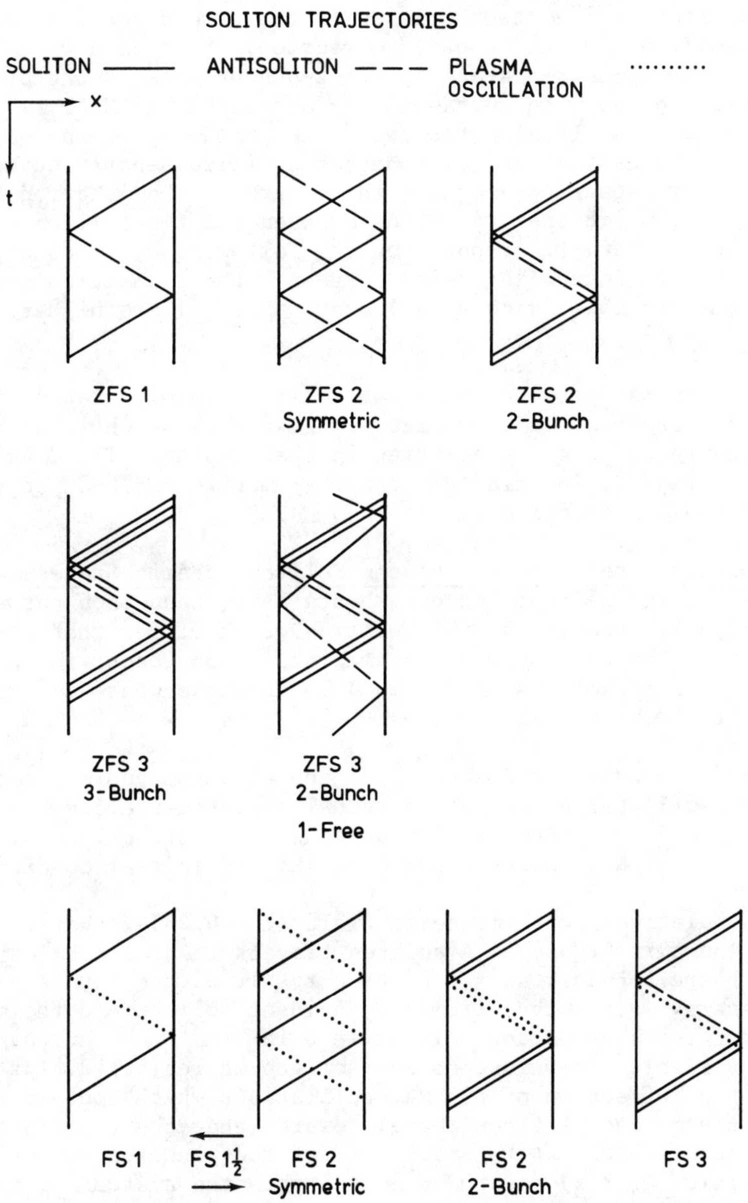

Figure 4. Soliton dynamic states. ZFS = zero field step corresponding to boundary conditions (4a). FS = Fiske step corresponding to boundary conditions (4b).

Figure 3 illustrates the 2-soliton case for different parameter values. For small values of γ (Fig. 3a) the two solitons travel in a symmetric configuration - i.e. soliton and antisoliton in opposite directions. For higher values of γ (Fig. 3b) the two solitons travel in a bunched state - i.e. two solitons in the same direction followed by two antisolitons in the opposite direction. There is a gradual transition from the symmetric state to the 2-bunch state as the bias current γ is increased and vice versa as γ is decreased. In Ref. [7] we have shown that the Hamiltonian for two (undeformed) solitons (on an infinite junction) has a local minimum for a finite separation between the solitons. This separation equals the length of the junction for the value of γ at which the transition between the two soliton dynamic states occurs. Ref. [2] reports on the following hysteresis phenomenon: For increasing (decreasing) bias current γ the transition from symmetric to bunched mode (vice versa) occurs at smaller (higher) values of γ.

We have investigated a number of soliton dynamic states recently. Some of the results are summarized in Figure 4, which illustrates the corresponding soliton trajectories in the xt-plane. The diagram ZFS 1 represents Fig. 2. The diagrams ZFS 2 symmetric and 2-bunch represent Figs. 3a and b respectively.

4. Chaotic Intermittency Between Soliton Dynamic States. In the previous section we reported on a hysteresis phenomenon between the soliton dynamic states. In this section we shall see that also chaotic intermittency between soliton dynamic states may occur for certain values of the parameters $(\alpha,\beta,\gamma,\ell,\eta)$ in the case where an external magnetic field is present $(\eta \neq 0)$.

In order to tune the frequency of the electromagnetic radiation from the oscillator a constant external magnetic field may be applied. The corresponding mathematical model then consists of the sine-Gordon equation (3) with boundary condition (4b) and initial condition (5).

For a relatively weak magnetic field ($\eta = 0.75$) we get the computer picture shown in Figure 5. A soliton travels towards the left and reacts with the boundary at $x = 0$. As a result of the boundary condition $\phi_x = \eta$ energy is absorbed from the incident soliton and the minimum energy for the sine-Gordon soliton (= 8 in normalized units) is no longer available. Therefore no antisoliton is reflected. Instead we observe the reflection of plasma oscillations which contain less energy. The plasma oscillations travel towards the other end in the figure and excite a soliton at the right end of the junction where the boundary condition $\phi_x(\ell,t) = \eta$ pumps energy into the system and makes this creation possible. This constitutes the first cycle of a stationary soliton dynamic state under the influence of a constant magnetic field. The state gives rise to a branch of the IV-characteristic for the oscillator designated the first Fiske step (FS 1 in Fig. 4).

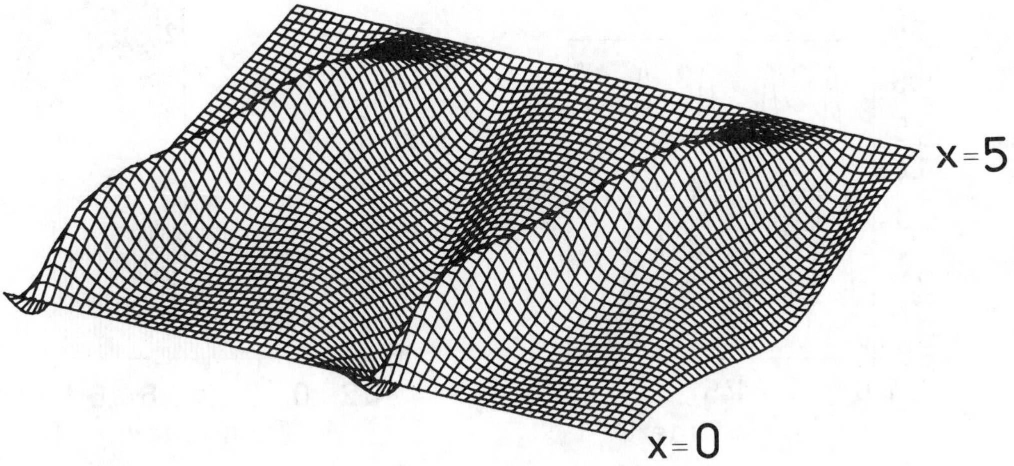

Figure 5. Computer solution of the perturbed sine-Gordon equation (3) with α = 0.252, β = 0, γ = 0.54. Boundary conditions (4b) with ℓ = 5, η = 0.75. Initial conditions (5) with one soliton.

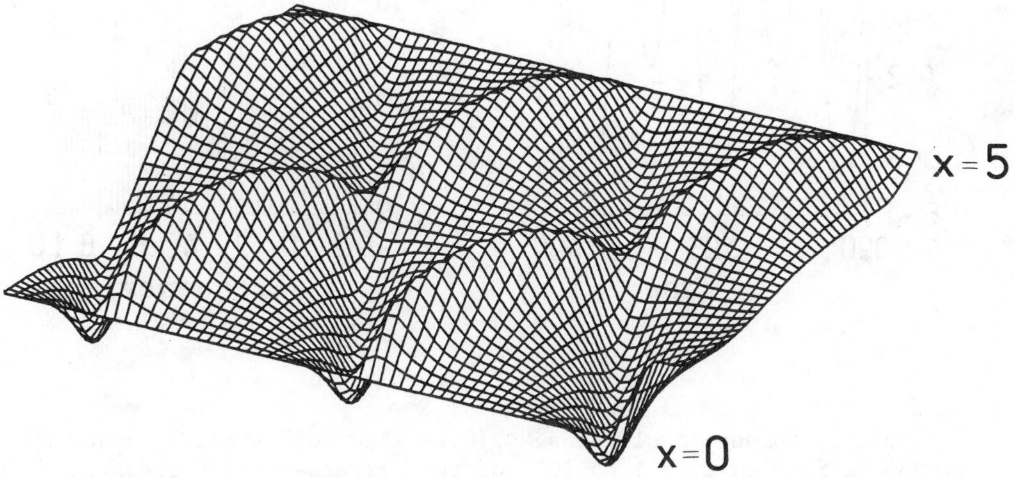

Figure 6. Computer solution of the perturbed sine-Gordon equation (3) with α = 0.252, β = 0, γ = 0.71. Boundary conditions (4b) with ℓ = 5, η = 1.35. Initial conditions (5) with two solitons.

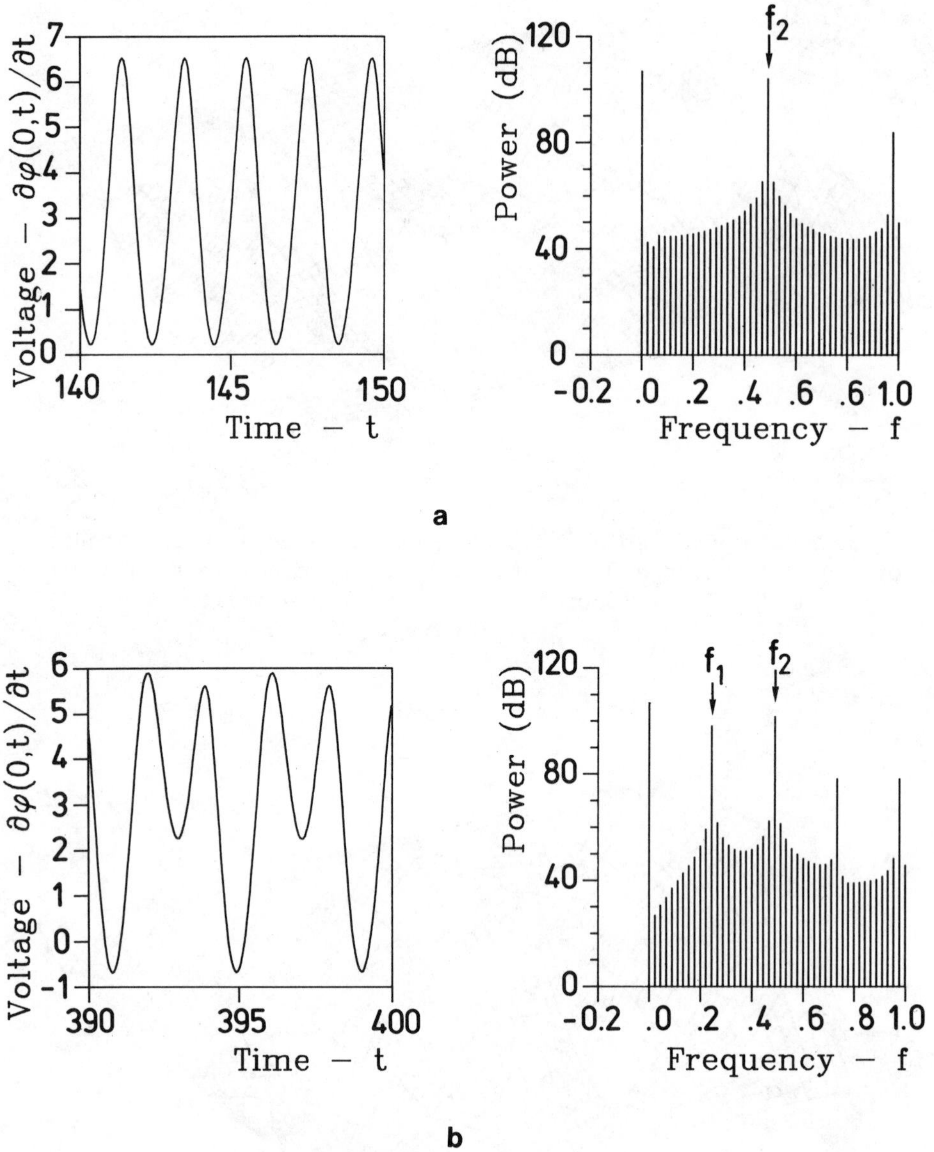

Figure 7. Computer solutions, $\phi_t(0,t)$, and corresponding power spectrum normalized to $\phi_t^2 = 1.0 \times 10^{-10}$ of the perturbed sine-Gordon equation (3) with $\alpha = 0.05$, $\beta = 0.02$. Boundary conditions (4b) with $\ell = 2$. Initial conditions (5) with two solitons. (a): $\gamma = 0.30$, $\eta = 2.5$. Symmetric configuration. (b): $\gamma = 0.28$, $\eta = 4.6$. Bunched configuration [8].

For higher values of η we may have two solitons present on the oscillator as illustrated in Figure 6. Here the soliton and the plasma oscillations travel in opposite directions at the same time in a symmetric configuration giving rise to a second Fiske step in the IV-characteristic (FS 2 symmetric in Fig. 4). Our recent work [8] has demonstrated that the solitons on the second Fiske step may also be bunched (FS 2 2-bunch in Fig. 4) when the surface loss parameter $\beta \neq 0$. This shift is similar to the shift from symmetric to bunched state on the second zero field step. The shift may be provoked for constant bias current γ by increasing the external magnetic field η. The shift is accompanied by characteristic changes in the power spectrum for the voltage at the end of the junction as illustrated in Figure 7. These changes can be verified experimentally [8].

For certain parameter values in a narrow portion of parameter space $(\alpha,\beta,\gamma,\ell,\eta)$ the oscillator switches back and forth between the two soliton dynamic states shown in Fig. 5 (FS 1) and Fig. 6 (FS 2) as demonstrated in Ref. [9].

The switching appears in the solution curve, $\phi(0,t)$, as a shift in the average slope of curve shown in Figure 8. Each little wriggle of

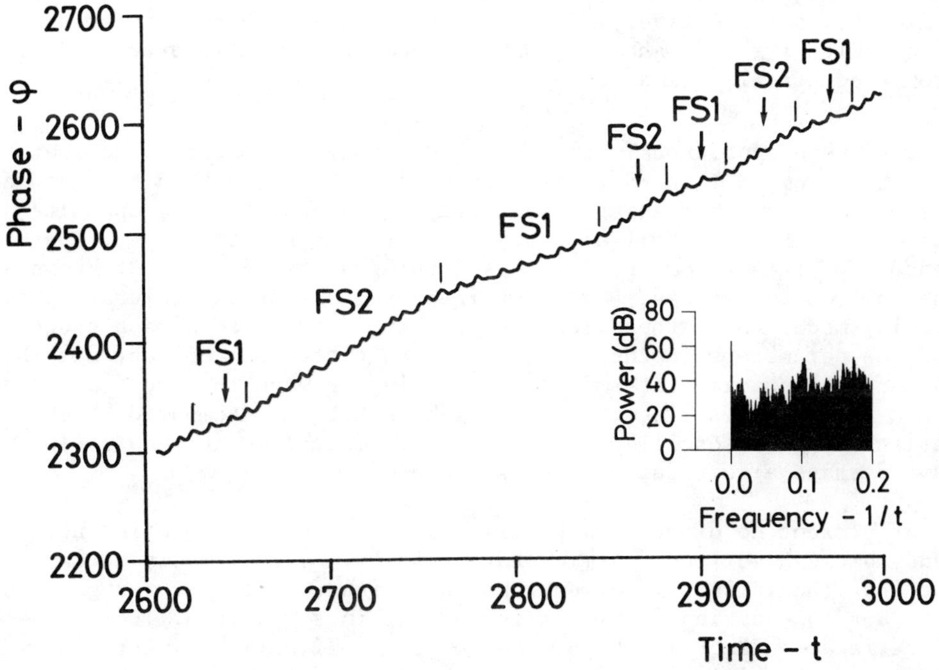

Figure 8. Computer solution, $\phi(0,t)$, and corresponding power spectrum normalized to $\phi_t^2 = 4 \times 10^{-7}$ of perturbed sine-Gordon equation (3) with $\alpha = 0.252$, $\beta = 0$, $\gamma = 0.480$. Boundary conditions (4b) with $\ell = 5$, $\eta = 1.25$. Initial conditions (5) with one or two solitons [9].

the curve marks the passage of a soliton through x = 0. The computational solution of (3), (4b), and (5) was continued for a very long time (t = 10,000). The resulting statistics is summarized in Figure 9.

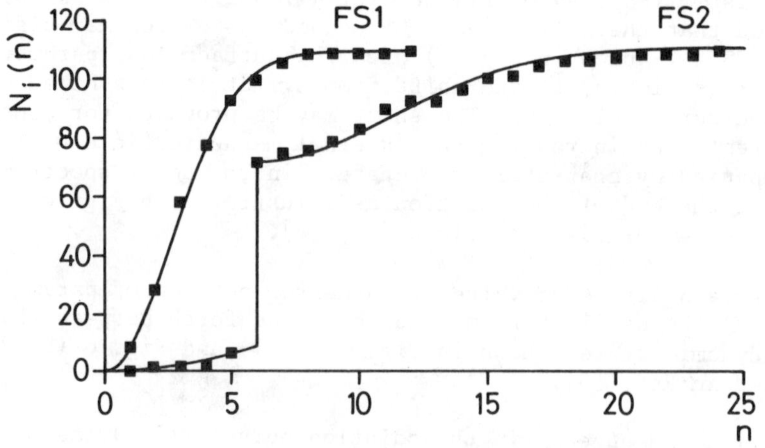

Figure 9. Abscissa: Length of time interval on FS 1 and FS 2 measured in terms of number of cycles, n, spent in the two soliton dynamic states respectively. Ordinate: Dots show number of intervals, $N_i(n)$, with i = 1,2, shorter than or equal to n. Full curves result from theoretical estimates [9].

We observe a continuously increasing probability that the oscillator switches from FS 1 to FS 2 as a function of time spent on FS 1 and a discontinuously increasing probability (at n = 6) for the opposite switch from FS 2 to FS 1 as a function of time spent on FS 2. The reason for this asymmetry in the switching statistics is not known. The full curves in Fig. 9 are obtained by means of an arbitrary probability model where the switching is assumed to be a Poisson process and the parameters in the model have been fitted in the best possible manner. The chaotic intermittent switching between the two soliton dynamic states gives rise to a special branch in the computed IV-characteristic for the oscillator. Recent experimental measurements [10] have perhaps also revealed such structures.

At present no exact or approximate theory for the chaotic intermittency between soliton dynamic states exists. Qualitatively we may argue that the infinitely dimensional non-linear system of the perturbed sine-Gordon equation is almost integrable in some sense. Therefore there is a tendency in the system to form soliton-like solutions as we have seen in the computer experiments. The soliton formation effectively requires infinitely many degrees of freedom and thus projects the system down to a non-linear low-dimensional system. This low-dimensional system exhibits chaotic intermittency between two critical points (corresponding to the two soliton dynamic states).

Acknowledgement. The financial support of the European Research Office of the Unites States Army through contract No. DAJA 37-82-C 0057 is gratefully acknowledged.

REFERENCES

[1] D.W. MCLAUGHLIN and A.C. SCOTT, Perturbation analysis of fluxon dynamics, Phys. Rev. A 18 (1978), pp. 1652-1680.

[2] P.S. LOMDAHL, O.H. SOERENSEN, and P.L. CHRISTIANSEN, Soliton excitations in Josephson tunnel junctions, Phys. Rev. B 25 (1982), pp. 5337-5748.

[3] M.P. SOERENSEN, P.L. CHRISTIANSEN, R.D. PARMENTIER, and O. SKOVGAARD, Subharmonic generation in Josephson junction fluxon oscillators biased on Fiske steps, Appl. Phys. Lett. 42 (1983), pp. 739-741.

[4] A.C. SCOTT, F.Y.F. CHU, and D.W. MCLAUGHLIN, The soliton: a new concept in applied science, Proc. IEEE, 61 (1973), pp. 1443-1483.

[5] P.L. CHRISTIANSEN, P.S. LOMDAHL, A.C. SCOTT, O.H. SOERENSEN, and J.C. EILBECK, Internal dynamics of long Josephson junction oscillators, Appl. Phys. Lett. 39 (1981), pp. 108-110.

[6] P.S. LOMDAHL, O.H. SOERENSEN, P.L. CHRISTIANSEN, A.C. SCOTT, and J.C. EILBECK, Multiple frequency generation by bunched solitons in Josephson tunnel junctions, Phys. Rev. B 24 (1981), pp. 7460-7462.

[7] T.H. SOERENSEN, P.L. CHRISTIANSEN, and P.S. LOMDAHL, An approximate theory for bunched solitons on finite Josephson tunnel junctions, Phys. Lett. 89 A (1982), pp. 308-309.

[8] M.P. SOERENSEN, R.D. PARMENTIER, P.L. CHRISTIANSEN, O. SKOVGAARD, B. DUEHOLM, E. JOERGENSEN, V.P. KOSHELETS, O.A. LEVRING, R. MONACO, J. MYGIND, N.F. PEDERSEN, and M.R. SAMUELSEN, Magnetic field dependence of microwave radiation in intermediate-length Josephson junctions, Phys. Rev. B (in press).

[9] M.P. SOERENSEN, N. ARLEY, P.L. CHRISTIANSEN, R.D. PARMENTIER, and O. SKOVGAARD, Intermittent switching between soliton dynamic states in a perturbed sine-Gordon model, Phys. Rev. Lett. 51 (1983), pp. 1919-1922.

[10] M. CIRILLO, G. COSTABILE, S. PACE, R.D. PARMENTIER, and B. SAVO, Possible observation of chaotic intermittency effect in a long Josephson junction in magnetic field, (to appear).

SYMMETRY AND COHERENCE IN A CHAIN OF WEAKLY COUPLED OSCILLATORS

NANCY KOPELL*

Abstract. We consider a weakly coupled chain of limit cycle oscillators with a weak gradient in freqency. It is shown that symmetry properties of the oscillators affect their ability to phase-lock; in particular, symmetry properties help determine the size of the gradient in frequency that can be sustained without loss of oscillator coherence. Also, when there is phase-locking, such properties affect the frequency at which the ensemble runs. For frequency gradients too large to allow phase-locking, local coherence remains after global coherence is lost.

1. Introduction. By an oscillator we shall mean a differential equation

$$\dot{X} = F(X) \tag{1}$$

with a stable limit cycle, where $X \in R^m$. As will be seen below, a collection of identical such oscillators, weakly coupled by nearest neighbor coupling, can "phase-lock". (For identical oscillators, this means simply that the homogeneous solution is asymptotically stable.) If, however, there is a variation in the frequencies of the oscillators, then the coupled system must balance two conflicting tendencies: the tendency to coherence encouraged by the coupling, and the tendency to dispersion produced by the variation in the underlying frequencies.

This paper, based on joint work with G.B. Ermentrout, has two themes. The first is that the above competition is influenced by certain symmetry properties of (1), and that this influence is strong

* Department of Mathematics, Northeastern University, Boston, MA 02115

for a long chain of oscillators. Secondly, even if the frequency variation is sufficiently strong to preclude total phase locking (defined below), one may get local "patches" of locking, i.e. local coherence. The mathematics described here was motivated by attempts to understand the phenomenon of "frequency plateaus" in mammalian intestine. For more information about the biological problem, and for references, see [1] and [2].

2. Derivation of the phase equations. We wish to study the coherence of chains of oscillators under conditions as general as possible. The main hypothesis is that the oscillators be uniformly close to one another. That is, the kth oscillator satisfies an equation of the form

$$(2) \qquad \dot{X}_k = F(X_k) + \varepsilon R_k(X_k, \varepsilon), \quad X_k \in R^m,$$

where, for $\varepsilon = 0$ (and hence for ε small), (2) has a stable limit cycle. (The reduction procedure to be outlined in this section actually holds under considerably less stringent conditions [3], but coherence fails if the oscillators are not uniformly close.) About the coupling, we assume that it is weak and linear. Thus, the full equations for the coupled system of N + 1 oscillators are

$$(3) \qquad \dot{X}_k = F(X_k) + \varepsilon R_k(X_k, \varepsilon) + \varepsilon D[X_{k+1} - 2X_k + X_{k+1}]$$

where D is an $m \times m$ positive definite matrix and $X_o \equiv 0 \equiv X_{N+2}$.

The phase space of (3) has dimension m(N+1). However, under the above hypotheses, all the relevant behavior takes place on an invariant (N+1)-dimensional torus. To see this, note that for $\varepsilon = 0$, there is a stable invariant torus for (3) which is the product of the (N+1) limit cycles of (2). The stability of those limit cycles implies that for ε sufficiently small, there is a nearby stable invariant torus for (3), [4]. (The bounds on ε are independent of the size of N because of the nearest-neighbor only coupling). This torus can be parameterized by $\theta_1, \dots, \theta_{N+1}$, where θ_k may be regarded as a phase for the limit cycle of (2). Using averaging techniques, it was shown in [1] that the equations on this torus may be written in the form

$$(4) \qquad \dot{\theta}_1 = \omega_1 + O(\varepsilon)$$

$$(5) \qquad \dot{\phi}_k = \varepsilon[\Delta_k + H(\phi_{k+1}) + H(-\phi_k) - H(\phi_k) - H(-\phi_{k-1})] + O(\varepsilon^2)$$

$$H(0) = 0 = H(\phi_{N+1})$$

Here $\phi_k \equiv \theta_{k+1} - \theta_k$, ω_k is the frequency of the limit cycle of (2) and $\varepsilon\Delta_k \equiv \omega_{k+1} - \omega_k$. H is a 2π-periodic function explicitly computable from (2). For similar computations in special cases, see [5], [6].

All the relevant properties of (2) and the coupling matrix D are now condensed into one scalar valued function H. For much of this paper, we shall be discussing the effects of a symmetry property of H (or lack thereof) on the behavior of (4), (5). For concreteness, we now include an example in which H is particularly simple.

Suppose that $X = (c_1, c_2)$, and

$$(6) \qquad F_k\begin{pmatrix} c_1 \\ c_2 \end{pmatrix} = \begin{pmatrix} \lambda & -\omega_k \\ \omega_k & \lambda \end{pmatrix}\begin{pmatrix} c_1 \\ c_2 \end{pmatrix} ; \; D = \begin{pmatrix} d_1 & d_2 \\ d_3 & d_4 \end{pmatrix}$$

where $\lambda, \omega: R \to R$, $\lambda(c_1^2+c_2^2) = 1-(c_1^2+c_2^2)$, $\omega_k = \overline{\omega}_k + \hat{\omega}[1-(c_1^2+c_2^2)]$. Equation (1), with F given by (6), has a stable limit cycle at $c_1^2 + c_2^2 = 1$. The frequency on this limit cycle is $\overline{\omega}_k$. $\hat{\omega}$ is a measure of the variation of local angular frequency with amplitude $c_1^2 + c_2^2$. If F and D are given by (6), then

$$H = A[1 - \cos\phi] + B \sin\phi$$

where

$$A = -\frac{1}{2}[(d_1+d_4)\hat{\omega} + (d_3-d_2)]$$

$$B = \frac{1}{2}[(d_2-d_3)\hat{\omega} + (d_1+d_4)]$$

3. <u>Phase-locking and symmetry</u>. We now return to the general case and look for solutions which are stable and coherent, i.e., ones in which all the oscillators are going at the same frequency. Note that, to lowest order in ε, (5) decouples from (4) and can be written

$$(7) \qquad \phi'_k = \Delta_k + H(\phi_{k+1}) + H(-\phi_k) - H(\phi_k) - H(-\phi_{k-1})$$

where $\dot{\phi}$ denotes d/dt, ϕ'_k denotes $d/d\tau$ and $\tau = \varepsilon t$. If the N dimensional system for (7) has a stable critical point, then the large systems (4), (5) or (2) have a stable limit cycle, which corresponds to a coherent periodic motion of the full system.

At such a critical point, the phase differences ϕ_k are independent of time (but not of k). Such a situation is known as phase-locking. As we will see below, even if the frequencies are not identical, phase-locking can occur, and the frequency differences are compensated for by phase lags ($\phi_k \neq 0$). For example, if the frequency decreases montonely down the chain, then the phase-locked situation all ϕ_k are negative. A given phase (e.g. $\theta = 2\pi\ell$ for some ℓ) then propagates down the chain as a wave (with non-uniform velocity). For a biological application of phase-locking in a chain of oscillators, see [7].

To determine if there is phase-locking, we must know the critical points of (7). We shall first assume that H is an odd function of ϕ, i.e. $H(-\phi) = -H(\phi)$. For concreteness, we shall write the following discussion for $H = \sin\phi$, but the conclusions hold for any odd function qualitatively like $\sin\phi$. The oddness of H greatly simplifies the calculation of the critical point. That is, with $\phi'_k \equiv 0$, (7) may be rewritten as

$$0 = \beta\underline{\Delta} + M\,\underline{\sin}\,\phi$$

where

$$\beta\underline{\Delta} = \begin{pmatrix} \Delta_1 \\ \vdots \\ \Delta_N \end{pmatrix}, \quad \underline{\sin}\,\phi = \begin{pmatrix} \sin\phi_1 \\ \vdots \\ \sin\phi_N \end{pmatrix}, \quad M = \begin{pmatrix} 2 & 1 & 0 & \cdot & \cdot \\ 1 & 2 & 1 & \cdot & \\ 0 & 1 & 2 & \cdot & \\ \cdot & \cdot & \cdot & \cdot & \end{pmatrix}.$$

(For a decreasing linear gradient in frequency, $\underline{\Delta} = -(1,\ldots,1)^t$, and β measures the strength of the gradient.) Since M is invertible, it is easy to solve for $\underline{\sin}\,\phi$:

$$\underline{\sin}\,\phi = -M^{-1}(\beta\underline{\Delta}) \tag{8}$$

From (8) it can be seen that if any component of $M^{-1}(\beta\underline{\Delta})$ has absolute value greater than one, then there are no critical points for (7). Thus there is a β_0 such that, for $\beta < \beta_0$, (8) has 2^N solutions, and for $\beta > \beta_0$, (8) has no solutions. Furthermore, it can be checked that $\beta_0 = O(1/N^2)$. It can also be checked that one of the critcial points for $\beta < \beta_0$ is stable [1].

We now contrast this with the behavior of (7) if H is not an odd function of ϕ. Again for concreteness, we use $H = [1-\cos\phi] + B\sin\phi$ (the H associated with (6), with A scaled to 1) and assume a linear gradient, but the ideas hold more generally [3]. It is no longer easy to compute the critical points of (7); indeed, we know of no explicit way to do this calculation. However, one can get information about the existence and properties of the critical points by noticing the analogy of (7) with a singularly perturbed partial differential equation. That is, (7) may be rewritten as

$$(9) \qquad \phi'_k = -\delta/N - [\cos \phi_{k+1} - \cos \phi_{k-1}] + B[\sin \phi_{k+1} - 2 \sin \phi_k + \sin \phi_{k-1}]$$

where $\delta/N = \beta$. This in turn may be written as

$$\phi'_k = \frac{1}{N} \{-\delta - \frac{[\cos \phi_{k+1} - \cos \phi_{k-1}]}{(1/N)} + \frac{B}{N} \frac{[\sin \phi_{k+1} - 2 \sin \phi_k + \sin \phi_{k-1}]}{(1/N^2)} \} .$$

The latter can be seen to be very similar to the partial differential equation

$$(10) \qquad \phi_\tau = \frac{1}{N} \{-\delta - [\cos \phi]_x + \frac{B}{N} [\sin \phi]_{xx} \}$$

where $0 \leq x \leq 1$ and $\phi_k(t) \approx \phi(x,t)$, $x = k/(N+1)$. At least for the time independent solutions, this analogy leads to a theorem:

Theorem: There is a δ_0 and $B_0 > 0$ such that, for $0 \leq \delta \leq \delta_0$, $B \geq B_0$ and N sufficiently large, (9) has a stable critical point. Furthermore, as $N \to \infty$, this critical point converges (non-uniformally as a function of k) to the time independent solution to (10) with boundary conditions $\phi(0) = 0 = \phi(1)$. (The latter corresponds to the "boundary conditions" on the system (9) that say that the first and last oscillators are each coupled at one end only. The non-uniformity is due to the singular nature of (10) as $N \to \infty$; there is a "boundary layer" in its time-independent solution.) For a proof, see [3].

It should be noted that, although the limit as $N \to \infty$ leads to an equation for a continuous variable, it is not at all the usual continuum limit. In the latter, one adds oscillators to the chain in such a way that the oscillators become arbitrarily close, and the local effect of diffusion becomes arbitrarily large; this would violate our assumption of weak coupling. Physically, the process expressed above corresponds to adding more elements to the chain, keeping the oscillators separated, and keeping the total frequency difference along the chain constant. (This is the point of the scaling $\beta = \delta/N$; δ/N is the (scaled with respect to ε) frequency difference between successive oscillators, so δ is the total (scaled) frequency difference.)

It may also be noted that the hypothesis $B \geq B_0$ is closely related to criteria for the numerical stability of discretized versions of singularly perturbed partial differential equations [8]. The term $(B/N)[\sin \phi]_{xx}$ can be regarded as a small "artificial viscosity" for a first order equation, and (9) as the discritezation of (10) into N spatial steps. B is a measure of the scale on which the viscosity works compared to the spatial step size. If B is too small, one expects spatial oscillations in solutions to (9) which are not reflections of solutions to (10). Indeed, numerical simulations of (9) for B small show time independent solutions which have such oscillations near the boundary layer. We emphasize that, for our application, the relevant equation is (9), with (10) just a "diagnostic" equation; these oscillations (in ϕ_k as k is varied) reflect real phenomena of coupled oscillators, not artifacts of an irrelevant discretization.

The above theorem allows us to read off some non-obvious facts. The first is that the ability to phase-lock is greatly enhanced for a system (3) whose H is not odd: as we have seen, for H odd, the maximal frequency difference in a phase-locked chain is $O(1/N^2)$. Thus the maximal total frequency difference along the chain is $O(1/N)$. This implies that for any fixed total frequency difference, for H odd and N sufficiently large, the system cannot phase-lock. By contrast, the theorem implies that for some non-odd functions H including $H = [1 - \cos\phi] + B \sin\phi$, the maximal total frequency difference in a phase-locked chain does not go to zero as $N \to \infty$. (It should be recalled that the above frequency differences are measured in the scaled time τ; in both the odd and non-odd cases, the frequency difference at which phase-locking is lost is, in the original time scale, $O(\varepsilon)$.)

The effect of the symmetry of H on the coherence of (3) can be understood physically, at least in the case of the model equations (6). H is an odd function whenever the coefficient A is zero. This coefficient has two terms, one of which reflects the dependence of the (local) frequency of each oscillator on the distance of a trajectory from the limit cycle; the other reflects the lack of symmetry in the diffusion matrix D. Assuming the latter is symmetric, a chain of oscillators whose frequency can change with amplitude can more easily accommodate to the frequency gradient.

The second non-obvious fact is that, if the system phase-locks at all, the symmetry properties affect the frequency at which the coherent system oscillates. For example, if $H = \sin\phi$ and $\underline{\Delta} = -(1,\ldots,1)^t$, it is easily seen by symmetry arguments that the frequency of the coherent system is the average frequency. By contrast, if $H = A[1 - \cos\phi] + B \sin\phi$, $B > 0$, and $\underline{\Delta}$ is as before, then for N large the frequency of the phase-locked system is very close to the highest or lowest frequency on the chain, depending on the sign of A[3]. This again can be rationalized physically, at least for the simple model (6): $A > 0$ corresponds to locking near the highest frequency. Now $A > 0$ means $\omega' < 0$ (if D is symmetric). Since coupling unequal oscillators tends to decrease their amplitudes, and $\omega' < 0$ implies that frequency increases with decreased amplitude, higher frequencies are favored.

4. Beyond phase-locking. Whether or not H is an odd function of ϕ, for sufficiently large gradients in the frequency of the underlying oscillators, coherence breaks down. However, the breakdown appears (at least initially) to be local.

Consider, for example, the case H odd (qualitatively like $H = \sin\phi$) and $\underline{\Delta} = -(1,\dots,1)^t$. As seen above, for $\beta > \beta_0$ (the critical value of the gradient), there are no critical points for (7), hence no phase-locking. It was shown in [1] that for $\beta < \beta_0$, (and $\beta_0 - \beta$ sufficiently small) there is a large amplitude stable invariant circle for (7) containing the stable critical point and one saddle; as $\beta \to \beta_0$, the critical points on this circle coalesce and give rise, for $\beta > \beta_0$, to a limit cycle for (7) (i.e. an invariant T^2 for (4), (5) or (3)). This limit cycle turns out to correspond to the existence of a pair of "frequency plateaus", i.e. stretches of oscillators for which the frequency is constant, with a break in frequency between the stretches. The position of the break is given by the homotopy class of the limit cycle in the N-dimensional toral phase space of (7), and for this symmetric case, the break is in the middle of the chain. (See [1] for details.) We note that, beyond phase-locking, θ_k is time-dependent, so "frequency" requires a definition. We are here using the average frequency θ_k over the limit cycle. It might also be pointed out that for $\beta - \beta_0$ small, the break in frequency varies like $\sqrt{\beta - \beta_0}$; thus there is no necessary rational relation between the frequencies on the pair of plateaus. For β larger (and H still odd, the gradient still linear), numerical experiements indicate that more frequency plateaus appear, but the number is not necessarily monotonic with β.

If H is not odd, we have seen that the phase-locking can occur for larger frequency gradients, and that the locked frequency is close to the highest or lowest frequency, rather than the average. Beyond phase-locking, this case is so far less understood. For sufficiently large gradients there again seems (numerically) to be a pair of plateaus corresponding to an invariant circle for (7); the break is close to the boundary layer of the phase-locked solution. For still larger gradients, more plateaus appear, and there is some evidence that the edges of the plateaus are not stable in time, suggesting temporal chaos. (See also [9].)

REFERENCES

[1] G.B. Ermentrout and N. Kopell, Frequency plateaus in a chain of weakly coupled oscillators, I, SIAM J. Math. Anal. 15 (1984), pp. 215-237.

[2] N. Kopell and G.B. Ermentrout, Coupled oscillators and mammalian small intestines, Lecture Notes in Biomathematics, 51, Oscillations in Mathematical Biology, pp. 24-36, J.P.E. Hodgson, ed., Springer-Verlag, N.Y., 1982.

[3] G.B. Ermentrout and N. Kopell, Symmetry and phase-locking in a chain of weakly coupled oscillators, in preparation.

[4] N. Fenichel, Persistence and smoothness of invariant manifolds for flows, Indiana Univ. Math. J., 21 (1971), pp. 193-226.

[5] R.H. Rand and P.J. Holmes, Bifurcation of periodic motions in two weakly coupled Van der Pol oscillators, Inter. J. Nonlinear Mech. 15 (1980), pp. 387-399.

[6] J. Neu, Coupled chemical oscillators, SIAM J. Applied Math., 37 (1979), pp. 307-315.

[7] A.H. Cohen, P.J. Holmes and R.H. Rand, The nature of the coupling between segmental oscillators of the lamprey spinal generator, J. Math. Biol., 13 (1982), pp. 345-369.

[8] A.Y. LeRoux, A numerical conception of entropy for quasi-linear equations, Math. of Comp., 31 (1977), pp. 848-872.

[9] B.H. Brown, H.L. Duthie, A.H. Horn and R.H. Smallwood, A linked oscillator model of electrical activity of human small intestine, Amer. J. Physiol., 229 (1975), pp. 384-388.

AN INFINITE DIMENSIONAL MAP FROM OPTICAL BISTABILITY WHOSE REGULAR AND CHAOTIC ATTRACTORS CONTAIN SOLITARY WAVES

D. W. McLAUGHLIN*, J. V. MOLONEY** AND A. C. NEWELL*

Abstract. In this article we summarize some of our recent results on coherence and chaos in optical bistable laser cavities. This project is important because of its specific application to optical logic elements and because of its general bearing on coherence and chaos in nonlinear wave equations. We describe and characterize the inevitable development of coherent transverse structures in the profile of the laser beam and the role of these structures in the subsequent routes of chaos.

I. Introduction. Optics has provided modern dynamicists with a rich variety of nonlinear phenomena: harmonic generation, self-focusing, self-induced transparency, simulated Brillouin and Raman scattering. The field is relatively young, having to await the invention and development of laser light in order that electric field intensities were large enough to cause matter to respond to light stimulation in a nonlinear manner.

Despite its late start, the field has progressed rapidly and has both given and taken new ideas to the general area of nonlinear science. Its latest point of contact concerns the notion of low dimensional chaos. This is a novel concept, developed in the early seventies, which provides an alternative scenario to the Landau picture for transition from a laminar to a turbulent state. The latter argued that "turbulence" was the end product of a large number of Hopf bifurcations which would take place as some stress parameter (such as the Reynolds number in fluids or light intensity in optics) was raised. It did not quite explain how a quasiperiodic motion with a discrete (althoughbeit eventually quite dense) spectrum evolved into a broad band continuous spectrum, but presumably if the states which the system can occupy are dense, a very small amount of noise would cause the system to drift in a nondeterministic fashion between the states.

---

*Department of Mathematics and Program in Applied Mathematics, University of Arizona, Tucson, AZ 85721

**Optical Sciences Center, University of Arizona, Tucson, AZ 85721

The authors are grateful for support to the Army Research Office (ARO Contract #DAAG29-81-K-0025), the Air Force of Scientific Research (AFOSR Contract #PR-83-00869), and the Office of Naval Research (ONR Contract #N00014-84-K-0420).

The new idea was quite revolutionary. It argued that the qualitative properties of the deterministic system were more important than the dimension. In particular, if the system had the property of sensitivity to initial conditions, then the dynamics would be such that orbits through neighboring points would diverge from each other at an exponential rate. As a result, in order to follow the trajectory of a phase point over ten time steps to within an accuracy of ten decimal places, one would need an initial accuracy of twenty decimal places. In other words, the dynamics erodes information and so even though the system is deterministic, its motion is unpredictable. Furthermore, this kind of behavior can be found in systems of ordinary differential equations of dimension three, in invertible maps of dimension two and even in one dimensional, noninvertible maps. Instead of the phase point being attracted (and here we are talking about dissipative systems) to a fixed point or a limit cycle, its motion ultimately takes place on a much more complicated set--a strange attractor--which, roughly speaking, for a dimension three system, is the direct product of a smooth surface and a Cantor set.

What this means in real physical situations is that chaotic behavior should be observable in systems whose dynamics are dominated by relatively few spatial modes. And this indeed turns out to be the case. There are many situations, usually at a relatively low value of the stress parameter, in which a potentially infinite dimensional system appears to be dominated by a few spatial mode structures but which nevertheless exhibits an aperiodic behavior in time. The most notable examples are the Taylor-Couette flow between rotating cylinders and Rayleigh-Bénard convection in small aspect ratio boxes. Indeed, this picture suggests an algorithm for handling these situations analytically. Simply project the field variable into a basis containing the dominant spatial structures and analyze the behavior of the low dimensional system of o.d.e.'s arising from this procedure. The difficulty is that it is very difficult to know what these modes are and how to use them as part of a basis of solutions. They are nonlinear and therefore do not allow direct superposition. An alternative procedure (Galerkin-like procedure) would be to pick a finite basis (like the first N Fourier modes) consistent with the geometry of the problem and project the field variables into this. Modes, created by nonlinear interaction, which do not belong, are simply ignored. This is called truncation. The difficulty with this scheme is that the behavior of the system is often most sensitive to N, the size of the basis. Therefore, in order to investigate infinite dimensional systems which appear to be dominated by a few spatial structures, it is crucial to find a description which allows one to capture the structure of these modes accurately.

Fortunately, there are examples of fully nonlinear problems in which the solutions can in some sense be written as a "superposition" of modes. These are the so-called soliton equations which are exactly integrable in the Hamiltonian sense and whose long time behavior is often dominated by a few soliton modes. One can now ask the following questions: Suppose such systems were perturbed. Can the resulting dynamics be described by simply looking at the resulting time evolution of the (previously fixed) soliton parameters? What other modes tend to arise? As the stress parameter is raised, what is the nature of the transition to temporal incoherence, to spatial incoherence? As we have noted, there is often a good reason for a system to retain a particular spatial structure beyond that value of the stress where temporal chaos first appears.

The model we have chosen to study in this paper arises in optically bistable cavities [1]. It is important for several reasons. First, it is of interest in its own right as bistable ring or Fabry-Perot cavities have been suggested as candidates for the optical logic element. Since optical transmission is already a reality, this would be an important step in the design of the first all optical computer. It is clear that the behavior of the logic element at all likely values of stress parameters should be studied. Second, it is accessible to experimental, numerical and analytical investigations. At the present time, with respect to behavior transverse to the propagation direction, the latter two approaches are ahead of the first, but a group in the Optical Sciences Center at Arizona is currently planning an experiment which can be compared with theoretical and numerical studies. Third, and perhaps most important from the point of view of nonlinear dynamics in general, the model provides us with an excellent vehicle for studying the bifurcations which lead to richer and more complicated spatial structures and for ultimately investigating the loss of spatial coherence altogether.

Our mathematical problem can be stated concisely. We study an infinite dimensional map

$$G_{n-1}(\cdot,\ell) \to G_n(\cdot,\ell) \tag{1.1}$$

which is defined by a sequence of initial value problems:

$$2i\frac{\partial}{\partial z} G_n + \frac{1}{f} \Delta G_n - \frac{G_n}{1+2G_n G_n^*} = 0, \tag{1.2a}$$

$$G_n(\vec{x},z=0) = a(\vec{x}) + Re^{ikL} G_{n-1}(\vec{x},z=\ell) \tag{1.2b}$$

where $n \geq 1$, $G_o=0$. Here f,R,kL are given parameters, Δ is the laplacian in $\vec{x}$, and $a(\vec{x})$ is a given function which is shaped like a Gaussian with maximum $a_o=a(o)$. Our goal is to find the behavior of the function $G_n(\vec{x},\ell)$ as $n\to\infty$.

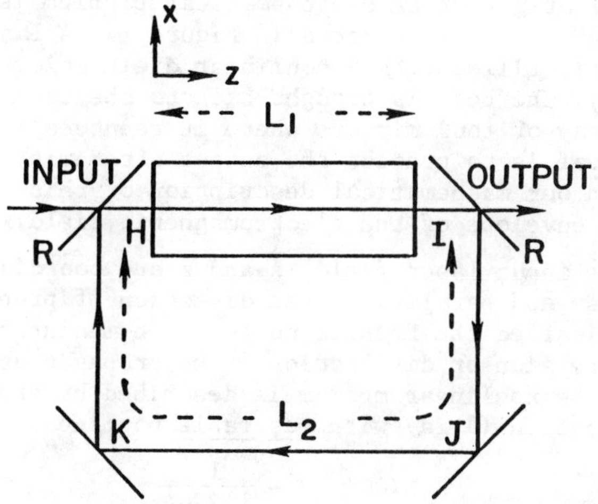

Figure 1. A schematic of an optical ring cavity.

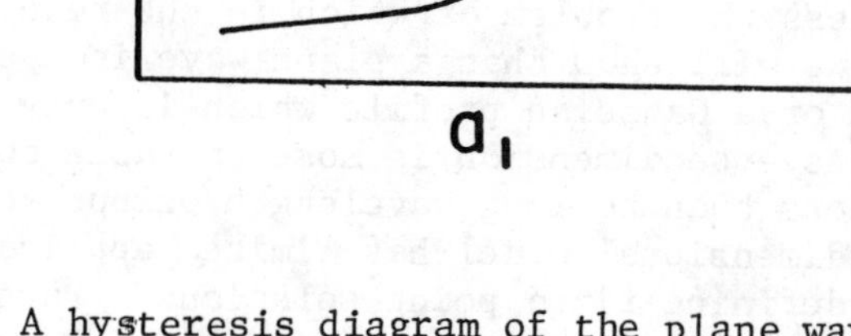

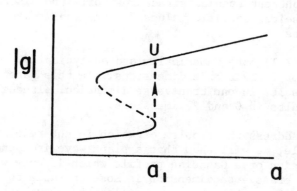

Figure 2. A hysteresis diagram of the plane wave map T. The output amplitude $|g|$ vs. the input amplitude a.

The physical origin of this mathematical problem is as follows: Consider a ring cavity as depicted in Figure 1. A laser beam enters a cavity which is filled with a nonlinear dielectric. As it emerges from the cavity, the beam is brought back to the entry point by a rectangular array of four mirrors where it reenters the nonlinear medium. The goal is to predict the output after many passes through the cavity. In our mathematical description of this process (1.2a,b), G_n denotes the envelope of the electromagnetic field, $a(\vec{x})$ is the envelope of the input laser field, $\vec{x}$ and z are coordinates in directions transverse and parallel to the direction of propagation. The parameter f is called the Fresnel number. It measures the amount of transverse dispersion or diffraction. The propagation of the laser envelope down the nonlinear medium is described by the nonlinear Schrödinger equation (1.2a) with <u>saturable</u> nonlinearity $N(GG^*)$,

$$N(GG^*) = -\frac{1}{1+2GG^*} . \tag{1.3}$$

The dynamics on the return path back to the entry point is accounted for by the factor Re^{ikL} which involves mirror losses ($R<1$) and a phase shift.

The infinite dimensional map (1.1) is the composition of two maps-one, (1.2b), is a dissipative discrete map which acts pointwise in $\vec{x}$ on the output $G_{n-1}(\vec{x},\ell)$ to produce the input $G_n(\vec{x},0)$, while the second, (1.2a), is a conservative nonlinear wave equation which transports the field down the nonlinear medium. In our analysis we will utilize separately properties of each of these two components of map (1.1). For example, the nonlinear wave equation component will filter the field into coherent spatial structures while the discrete dissipative map will select specific values for the amplitudes of these spatial structures.

We study map (1.1) both numerically and analytically. Our results differ in 0, 1, or 2 transverse dimensions. In this paper we will emphasize the results in one transverse dimension, although we will mention some results in 0 and 2 dimensions.

We will also address the problem of which is the relevant dimension. For example, we will show that a plane wave independent of the transverse dimension or a Gaussian profile which is only weakly dependent on the transverse dimension is more unstable to shortwave transverse fluctuations than to long wavelength perturbations. This means that the zero dimensional model has limited applicability for anything other than defining fixed point solutions. Pursuing this notion further, one might argue that, in turn, two transverse dimensional fields (because of the self-focusing effect) are more probable than one dimensional ones. We show that this is not always the case; often the most important solutions (which are of the one dimensional category) are cylindrical ring structures.

In each dimension, the numerical experiments may be summarized by describing the bifurcation sequence which is recorded as a stress parameter, such as a_o, is increased. For moderate values of a_o, the output $G_n(x,\ell)$ approaches fixed points as n increases. Frequently, multiple fixed points coexist. In particular, one often has three as in the common "hysteresis diagram" of Figure 2, with the upper and lower branches stable and the intermediate branch unstable to like excitations. For larger values of a_o, this fixed point becomes unstable and $G_n(x,\ell)$ approaches a period 2 (in n) attractor. At still larger values of a_o, $G_n(x,\ell)$ becomes chaotic in n. In all cases these attractors (fixed point, period two, chaotic) appear to be spatially coherent.

In the remainder of this introduction we will list our main results. (Each will be discussed somewhat in the text, and complete details will appear in references [2, 3, 4].) (i) The central part of the transverse profile of the typical "upper branch" fixed point $G_*(x,\ell)$, and also of the period 2 and chaotic attractors, is accurately described by solitary waves. (ii) The wings of this transverse profile have a different structure. For example, the wings of the fixed point $G_*(x,\ell)$ consist of a flat plateau. This plateau can be removed by altering the input field a(x); its presence or absence drastically effects the bifurcation sequence which occurs. (iii) Using a solitary wave basis, one can reduce the infinite dimensional map (1.1) to a two dimensional real map which very accurately predicts the amplitude of the central portion of the fixed point $G_*(x\simeq 0,\ell)$. Alternatively, the two dimensional, real "plane-wave" map of Ikeda [5] accurately predicts the height of the flat plateau in the wings. (iv) Each of these two reduced maps experiences period doubling bifurcations, and the Ikeda plane wave map possesses the classic period doubling route to chaos of Feigenbaum [5, 6]. Nevertheless, when a flat plateau is present in the wings of G_*, the system chooses an entirely different route to chaos than predicted by either of these two reduced maps. In fact, we find that the plane wave map, while useful in orienting our studies, has very little to do with reality when transverse degrees of freedom are present no matter how large the Fresnel number. When the central part of the beam switches to the upper branch, large spatial gradients appear which force diffractive effects [6]. Even on the lower branch we find the dominant instabilities have spatial ripples; they are not flat. Transverse structure seems inevitable! (v) As the stress parameter a_o is increased, the flat plateau fixed point becomes unstable to a period two state whose spatial structure consists of ripples of a fixed wavelength. Using the full infinite dimensional map, we describe this period doubling bifurcation analytically--computing the spatial wavelength and threshold intensity. This phenomena and its mathematical description are new and should be widely applicable in other situations. (vi) As a_o is further increased, these spatial ripples interact with the mean of the wings and produce chaos. This route to chaos involves several degrees of freedom and thus differs fundamentally from those routes described by the two reduced maps. (vii) Solitary waves inhibit this chaos. They seem to ride with the chaotic background rather than actively participate in the chaos. We will describe a very beautiful numerical experiment which will show, in a controlled manner, the nature of the bifurcation sequence in the presence and absence of the solitary waves. (viii) Finally, in two transverse dimensions, both rings and filaments can describe the asymptotic states. We have seen both in numerical experiments, and have done a stability calculation which indicates which structure will be seen in given circumstances.

II. The Plane Wave and Solitary Wave Reduction

A. The Plane Wave Case. Our analytical study of the infinite dimensional map (1.1) proceeds by reduction in its dimension to a tractable size. In this paper we describe three different reductions. The first method simply ignores transverse variation by assuming that the incident laser pulse $a(\vec{x})$ is independent of $\vec{x}$. Thus, in this "plane wave" case, the output fields G_n are also independent of $\vec{x}$ and differential equation (1.2a) can be solved explicitly,

$$g_n(z) = \exp\left[\frac{i}{2}N(g_n g_n^*)z\right]g_n(0). \qquad (2.1)$$

Here $G_n(\vec{x},z)=g_n(z)$, $g_n g_n^*$ is independent of z, and $N(gg^*)=-(1+2gg^*)^{-1}$. Inserting this solution (2.1) into the map (1.2b) yields the plane wave map T [5]:

$$g_{n+1} = T(g_n,g_n^*) \equiv a + R\exp\left[i\left(kL + N(g_n g_n^*)\frac{\ell}{2}\right)\right]g_n \qquad (2.2)$$

Plane wave map (2.2) is a one dimensional complex, or a two dimensional real, invertible map. Since it depends upon $g_n g_n^*$, it is not analytic in g_n. This map $T(\cdot)$ works as follows (Figure 3): First, the point $g=g_n$ is rotated through an angle $\theta = (kL + \ell N(gg^*)/2)$ which depends nonlinearly upon g; then the new point is moved along the ray toward the origin by a contraction factor $R<1$; finally, the point is shifted to the right by a positive real number a.

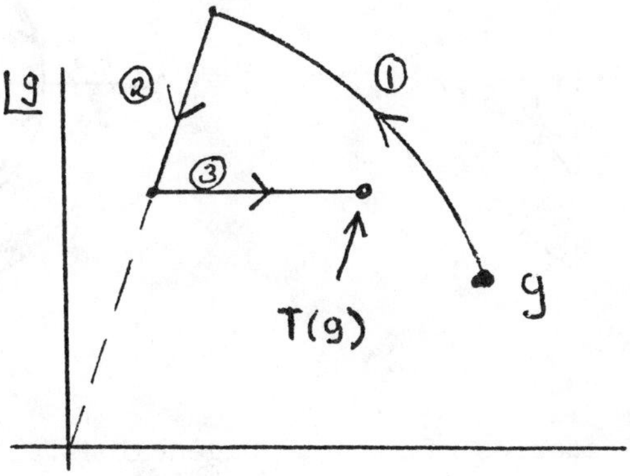

Figure3. The action of the plane wave map T on a (complex) point g.

This geometrical description of the action of map $T(\cdot)$ makes the existence of fixed points plausible. In fact, they do exist.

A fixed point g satisfies

$$g = T(g,g^*) = a + R \exp[i(kL + N(gg^*) \frac{\ell}{2})]g. \qquad (2.3)$$

In order to obtain quantitative information about these fixed points, this transcendental equation must be solved numerically [6]. However, analysis does yield qualitative information [5,6,7,8]. For example, (i) the parameter ℓ, a dimensionless measure of the length of the nonlinear medium, primarily determines the number of fixed points which increases as ℓ increases. (ii) In the situation depicted in Figure 2, there are three fixed points, and the fixed points as function of a take on the familiar "hysteresis shape" of Figure 2.

The stability of these fixed points can be studied analytically [5, 7, 8]. Consider the map $T(\cdot,\cdot)$ and linearize it about a fixed point (g,g^*):

$$g + \tilde{h} = T(g + \tilde{g}, g + \tilde{g}^*) \Rightarrow$$

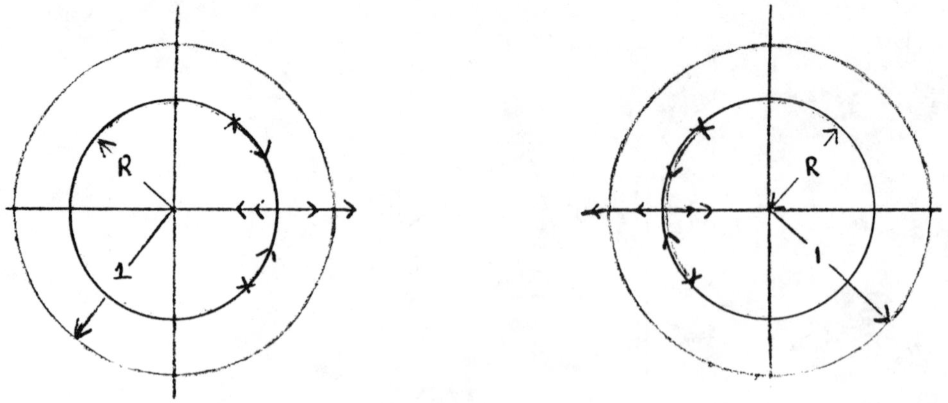

Figure 4. Trajectories of the μ eigenvalues.

$$\begin{pmatrix} \tilde{h} \\ \tilde{h}^* \end{pmatrix} = D_g T \begin{pmatrix} \tilde{g} \\ \tilde{g}^* \end{pmatrix} , \tag{2.4a}$$

where the 2x2 matrix $D_g T$ is defined by

$$D_g T \equiv \begin{bmatrix} (1 + i\frac{\ell}{2} N'(I)I)Re^{i\Gamma} & i\frac{\ell}{2}N'(I)g^2 e^{i\Gamma} \\ -i\frac{\ell}{2} N'(I)g^{*2}e^{-i\Gamma} & (1 - i\frac{\ell}{2}N'(I)I)Re^{-i\Gamma} \end{bmatrix} \tag{2.4b}$$

Here $I = gg^*$ and $\Gamma = \Gamma(I) \equiv kL + N(I)\ell$. Notice $\det D_g T = R^2 < 1$. The eigenvalues (μ_1, μ_2) of $D_g T$ satisfy

$$\mu_1\mu_2 = R^2 < 1$$

$$\mu_1 + \mu_2 = (1 + i\frac{\ell}{2} N'(I)I)Re^{i\Gamma} + (1 - i\frac{\ell}{2} N'(I)I)Re^{-i\Gamma}. \tag{2.5}$$

The fixed point g is linearly stable to forward iterations if and only if $|\mu_j| < 1$ for $j = 1,2$. Here two cases arise--(i) μ_1 and μ_2 both real or (ii) $\mu_2 = \mu_1^*$. The second case of conjugate pairs is always stable, since $\mu_1\mu_2 = \mu_1\mu_1^* = R^2 < 1$. Thus, no "Hopf bifurcations" are allowed. In the first case of real eigenvalues, either both satisfy $|\mu_j| < 1$ or one satisfies it and the other does not. As an eigenvalue μ crosses $+1$, the instability which occurs is a "saddle-node" bifurcation. As an eigenvalue μ crosses -1, a period 2 bifurcation occurs. Typical trajectories of the eigenvalues as the parameters change are depicted in Figure

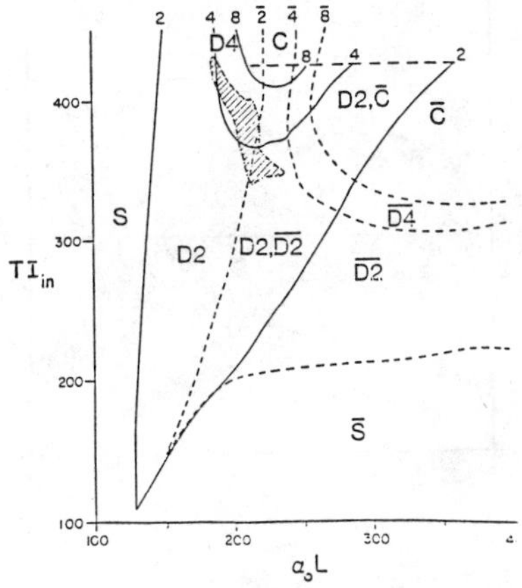

Figure 5. A bifurcation diagram for the plane wave map T. In the diagram, the horizontal axis is a measure of the cavity length, while the vertical axis measures the input intensity. S and $\bar{S}$ label fixed points; D2 and $\overline{D2}$ period 2 states; D4 and $\overline{D4}$ period 4 states; C and $\bar{C}$ chaotic states.

This linearized stability analysis can be applied to hysteresis loops such as Figure 2. Frequently one finds that the upper and lower branches are stable, while the intermediate branch is unstable. Thus, we have the characteristics of a yes-no optical switch. For the saturable nonlinearity, it is rare that the upper branch becomes unstable. However, the lower branch (or intermediate branches when there are multiple (5,7) fixed points) can become unstable, for example, to a period 2 bifurcation.

Rather than continue the detailed algebraic analysis of (2.5) in order to study the linearized stability, we numerically study the behavior of T(·) after many iterations. These studies are summarized in Figure 5, which shows a bifurcation diagram as a function of parameters kL and a^2. This diagram shows stable fixed points going unstable to period 2 states, which in turn go unstable through a period doubling route to chaos. It also shows that different chaotic states (with different basins of attraction) can coexist at the same parameter values.

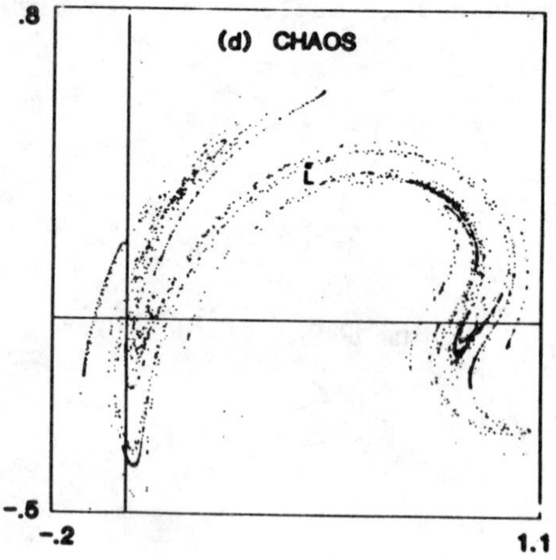

Figure 6. Iterates of the map for T in a chaotic region of parameter space, after transients have died out.

Numerical experiments can be used to study these chaotic attractors. First, a sequence of iterates can be plotted on the complex phase plane, Figure 6. This figure shows the leafy-Cantor like structure which is familiar in "strange attractors".

Using global phase space methods [8], considerably more information can be obtained about the map T(·). In particular, it is important to understand the universal, self-similar properties of this 2-dimensional, invertible map in a class of maps which are the composition of a nonlinear rotation, a contraction, and a shift. However, for our purposes the more elementary analysis summarized here is sufficient.

We close this section on the plane wave map by returning the reader's attention to its hysteresis diagram Figure 2, which will play an important role in the next section.

II.B. Solitary Wave Reduction. We return to the original problem (1.1) and include one dimensional transverse effects by considering an input field $a(x)$ with a Gaussian-like transverse profile,

$$2i\frac{\partial}{\partial z}G_n + \frac{1}{f}\frac{\partial^2}{\partial x^2}G_n + N(G_nG_n^*)G_n = 0$$

$$G_n(x,0) = a(x) + Re^{ikL}G_{n-1}(x,\ell) \tag{2.6a,b}$$

Clearly this infinite dimensional map possesses an extremely wide variety of potential responses depending upon parameter values. We restrict these by focusing our attention on large Fresnel numbers ($f \simeq 200 >> 1$) and selecting the parameters kL, ℓ, and R in regions near which the plane wave map has a hysteresis diagram such as Figure 2.

In equation (2.6) the only coupling of a transverse segment of the beam profile to its neighbors occurs through the laplacian $f^{-1}\partial_{xx}$. For large Fresnel number, $f^{-1}\partial_{xx}a(x) << a(x)$ this coupling can be neglected initially. Then each transverse segment of the profile acts independently from its neighbors according to a local plane wave theory. Thus, those points of the Gaussian profile for which $a(x) > a(x_\pm) \equiv \sqrt{I_1}$ will switch up to the upper branch while those points for which $a(x) < a(x_\pm) \equiv \sqrt{I_1}$ will go to the lower branch. The center of the beam profile will switch up, while the wings will not. For the saturable nonlinearity, this situation is shown in Figure 7. There we show the initial Gaussian profile and the output profile after 23 passes through the nonlinear medium. Notice that the center of the profile has switched to the upper branch while its wings have switched to the lower branch according to the plane wave theory.

However, now the two outer edges of the profile possess a steep gradient and, near $x \simeq x_\pm$, $f^{-1}\partial_{xx}G_n$ is no longer negligible. The plane wave approximation is no longer valid. No matter how large the Fresnel number f, transverse effects must be taken into account. Numerical experiments (which solve the full partial differential equation) show what happens during this stage of the evolution. At $n \simeq 30$ narrow pulses of width $\Delta x = O(1/\sqrt{f})$ are generated (Figure 8). These narrow pulses eventually fill out the region between x_- and x_+. The number of pulses which eventually fill out the transverse profile can be controlled. For the saturable nonlinearity, we have observed

1,3,5,7 in different situations. This number of pulses seems a function of the transient shape realized after approximately 20 passes; it is proportional to $\sqrt{f}$ and to the maximum input field amplitude.

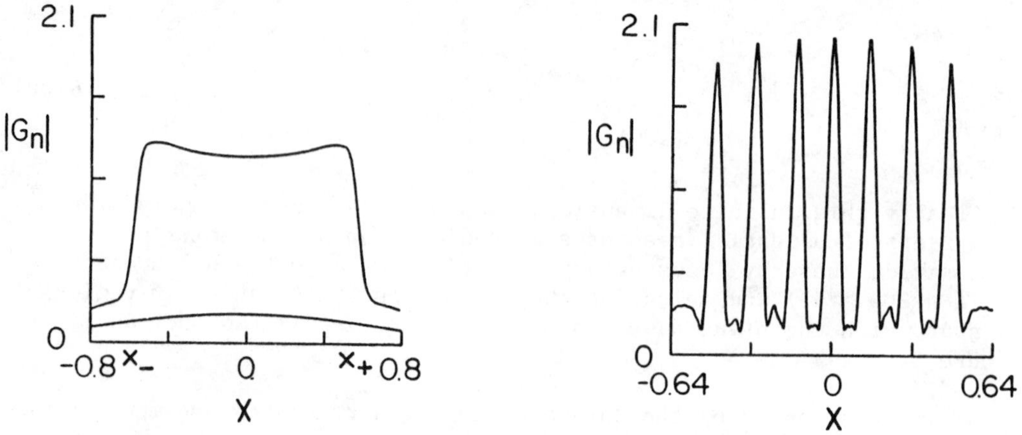

Figure 7. a) The inital Gaussian profile and the output profile after 23 passes through the medium. b) The asymptotic profile (in n) of the same experiment.

Once these spatial pulses form and fill up the transverse profile, they persist and describe the large n asymptotic response of the map. These transverse pulses can become the steady-state response of the system [2]; that is, they are stable fixed points of the infinite dimensional map (1.2). These fixed points can, as a_o is increased, become unstable to a period 2 state, which in turn becomes temporally (in n) unstable to a chaotic state [3]. We emphasize that as these bifurcations occur, the spatial coherence of these transverse pulses persists to a high degree. As yet, we have not observed any spatial chaos. Before describing our numerical results which show the spatial coherent pulses undergoing bifurcations from temporal fixed points through temporal period 2 into a temporally chaotic state, we describe the simplest case when the transverse pulses are stable fixed points of the infinite dimensional map (1.1).

Known theory for the nonlinear wave equation (2.6a) indicates that the transverse pulses should be solitary waves. A solitary wave is a solution of (2.6a) in the form ($y = \sqrt{f}\, x$)

$$G_s(y,z;\lambda,\gamma) = S(\lambda y;\lambda)e^{i[(\lambda^2-1)z/2 + \gamma]}, \qquad (2.7a)$$

where $S(\theta;\lambda)$ is a real, even solution of

$$S_{\theta\theta} - S + \frac{1}{\lambda^2}\,[1 + N(S^2)]\; S = 0 \tag{2.7b}$$

which vanishes as $\theta \to \infty$. The solitary wave (2.7a) is a two-parameter family of solutions of (2.6a) with parameters λ, γ which are free to be chosen. λ determines the amplitude and width of the solitary wave, while γ determines its phase. (In the case of the Kerr nonlinearity $[1 + N(S^2) = 2\, S^2]$, the solitary wave takes the explicit form

$$S(\theta;\lambda) = \lambda \operatorname{sech} \theta. \tag{2.8}$$

In this case λ clearly determines the amplitude of $S(\theta;\lambda)$.)

For the saturable case in parameter regions where the transverse profile is a fixed point of (1.1), we checked numerically [2] that the profile is accurately described by solitary waves--at least for the central portion of the profile. These numerical checks will be described later. Here we remark that a good fit was anticipated on the basis of known theory for equation (2.6a). However, this conservative wave equation could not pick the amplitude of the asymptotic solitary waves. Usually the amplitude is picked by the initial data. What is particularly new in this study is that this amplitude is picked by stable fixed points associated with map (2.6b).

In order to describe this situation analytically, we use solitary wave perturbation theory to reduce the dimension of the infinite dimensional map (1.1) to a two-dimensional map on solitary wave parameters (λ,γ) [2,4]:

$$(\lambda_n, \gamma_n) \mapsto (\lambda_{n+1}, \gamma_{n+1}). \tag{2.9}$$

This map is given, implicitly, by

$$(S_{n+1}, S_{n+1}) = (A_{n+1}, S_{n+1}) \cos\gamma_{n+1} + R\cos\Gamma_{n,n+1} (S_{n+1}, S_{n,n+1})$$

$$0 = -(A_{n+1}, \rho_{n+1}) \sin\gamma_{n+1} + R\sin\Gamma_{n,n+1} (\rho_{n+1}, S_{n,n+1}) \tag{2.10}$$

where

$$A_n \equiv a(\theta/\lambda_n)$$

$$S_n \equiv S(\theta,\lambda_n)$$

$$\Gamma_{n,n+1} \equiv kL + (\gamma_n - \gamma_{n+1}) + \frac{\ell}{2}\,(\lambda_n^2 - 1)$$

$$S_{n,n+1} \equiv S(\lambda_n \theta/\lambda_{n+1}; \lambda_n)$$

$$\rho_n \equiv \frac{1}{\lambda_n} \theta S_\theta(\theta;\lambda_n) + \frac{\partial}{\partial\lambda} S(\theta,\lambda)\Big|_{\lambda = \lambda_n}$$

We have studied this reduced two-dimensional map both analytically and numerically [2,3]. The fixed points of this map predict the amplitude of the solitary waves which finally emerge after many passes through the nonlinear medium. For the saturable nonlinearity, we solved for the fixed points numerically and compared our predicted solitary wave shapes and amplitudes with those obtained by numerical experiment. Typical results are summarized in Figure 9. They are extremely accurate.

For the cubic Kerr nonlinearity, much more can be done analytically. For example, we show that the reduced map on solitary wave parameters can undergo a period doubling bifurcation, but only at very large values of the stress parameter a_0. In the next section, we will discuss what actually happens for the infinite dimensional map.

III. Summary of Recent Results

In this last section we briefly summarize some very recent unpublished results which we have obtained since completing the work described above. (i) Near the parameter range where the fixed point is a single solitary wave, we have observed numerically a period doubling (Figure 10). In each of the two states in the 2 cycle, the central portion of the transverse profile remains a solitary wave. Only its amplitude (and width) oscillates in a period 2 fashion. (ii) While the central part of the transverse profile of the fixed point is a solitary wave, its wings do not go to zero, but rather to a fixed height which is the lower branch of the plane wave hysteresis curve. For the period 2 state, this plateau has developed spatial oscillations (Figure 10a,b). (iii) As one moves in parameter a^2 farther from the fixed point, a period 2 state remains, but the shorter and broader of the profiles in the 2 cycle develops a "dimple" at its central peak, (Figure 10c). At the same time the spatial oscillations in its wings have become distinct, large amplitude oscillations. Thus, these period two states could not be pure solitary waves. (iv) As one moves a^2 still further, the response goes chaotic in n, (Figure 10d). The transition seems to go from period 2 with a dimple directly into chaos. The transverse spatial profiles seem very coherent, with the same basic structure of a solitary wave with dimple and large spatial oscillations in the wings. The chaos is rather mild, consisting basically of a modulation of the period 2 state. The chaotic state tries to remain period 2, but misses slightly. Then it experiences short times, which occur very irregularly (randomly?), over which it is nearly a single solitary wave fixed point. This "near fixed point" is temporally unstable, and the state quickly returns to a "nearly period 2" state. (v) We have done some diagnostics of this "chaotic attractor". For example, using the algorithm of Grassberger and Procaccia [9,10], we have the preliminary estimate on the dimension of this attractor, ~ 2.8. This is certainly low dimensional for an infinite dimensional system, and is consistent with the importance of only a few spatially coherent modes.

The spatial oscillations in the wings of the transverse profile play a central role in the development of the temporally chaotic state. Since the wings of the profile are closely related to the lower branch of the plane wave hysteresis diagram, we study a related problem which is entirely on the lower branch (with no solitary wave in the central portion of the profile). This situation is initiated by a much smaller stress parameter a_o. The bifurcation sequence for this lower branch phenomena is plane wave fixed point $\rightarrow$ period 2 state with definite spatial oscillations in its wings $\rightarrow$ a chaotic state with definite spatial oscillations. This sequence is depicted in Figures (11,12), together with a similar sequence at larger a_o where a solitary wave is excited.

Several points must be emphasized: (i) The bifurcation sequences are very similar whether the solitary wave is present or not. In particular, the chaotic state in the wings of the profile with solitary wave at its center is very similar to the lower branch chaotic state with no solitary wave. The main difference is that the presence of the solitary wave inhibits the chaos. The solitary wave seems to ride with the chaotic background rather than actively participate in the chaos. (ii) The onset of the chaos is not through the universal period doubling cascade of Feigenbaum. Rather, it goes from a fixed point, through a period two state with more spatial degrees of freedom, directly into chaos. (iii) The period doubling transition and the transition into chaos occur at smaller values of the stress parameter a_o than predicted by the plane wave theory, and at much smaller values than predicted from the reduced solitary wave map. (iv) Apparently, the extra degrees of freedom in the wings play an active role. This conjecture was checked by altering the Gaussian input field a(x). We find that by substantially reducing the width of the input profile, we can remove the wings from the asymptotic state and leave a rather pure solitary wave profile. The period doubling transition of this state occurs at much higher values of the stress parameter, which is consistent with the interpretation of the active role of the wings of the profile and with [4].

To analyze the structure of the wings [4], we return to the lower branch case and study the stability of the lower branch fixed point of the plane wave map with respect to full infinite dimensional perturbations. That is, we study the stability of this fixed point as a fixed point of the full infinite dimensional map (1.1). We find the first instability (the instability which occurs at lowest value of the stress parameter a_o) has precisely the spatial wavelength measured by the numerical experiments. In addition, we accurately predict the critical intensity and growth rate of the instability. No matter how large the Fresnel number, transverse spatial structure is inevitable even on the lower branch!

Finally, we have begun to study the map (1.1) in two transverse dimensions. In this situation the transverse structures which emerge after the initial switch up are concentric rings about the axis of the nonlinear medium. In the Kerr case, these rings are strikingly unstable to the formation of filaments, (Figure 13). They are much more stable in the saturable case, (Figure 14). We have developed an analytical stability theory which contrasts the instability of rings in the Kerr medium with their stability in the saturable medium [12].

Clearly many questions remain to be answered. One of the most important concerns the development of the chaotic state. In order to achieve this state, the finite amplitude oscillations in the wings must play a fundamental role. At present we are initiating an analytical description of this state.

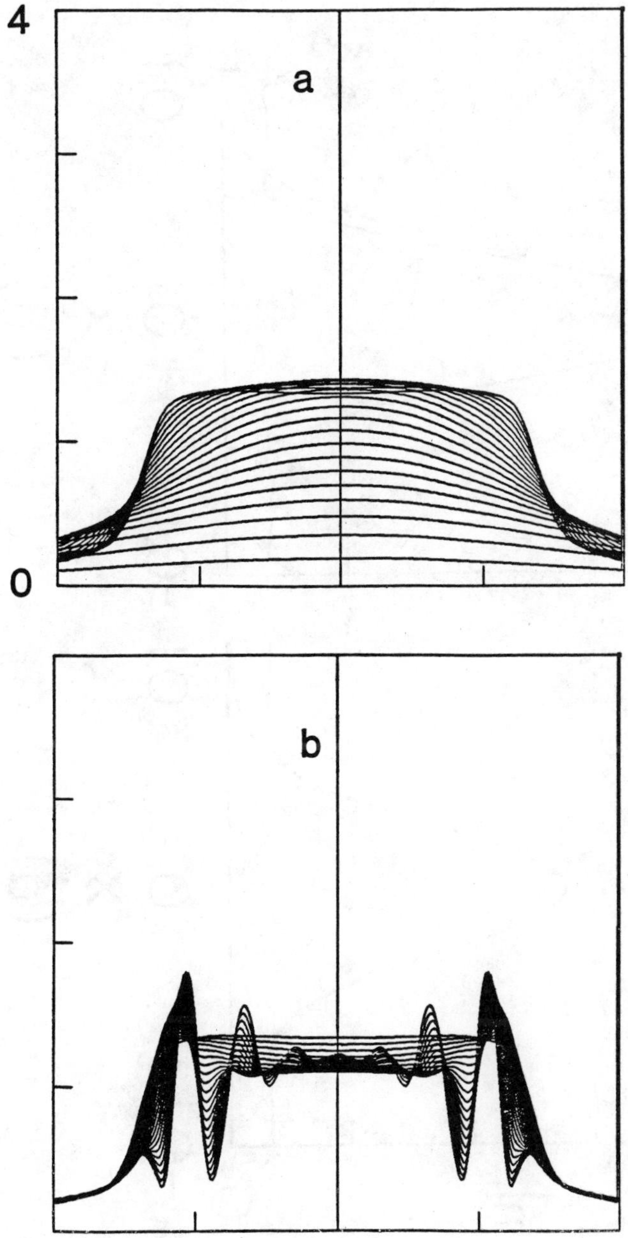

Figure 8. a) The switch up of transverse profiles during the first 20 passes. b) The formation of pulses in the transverse profile.

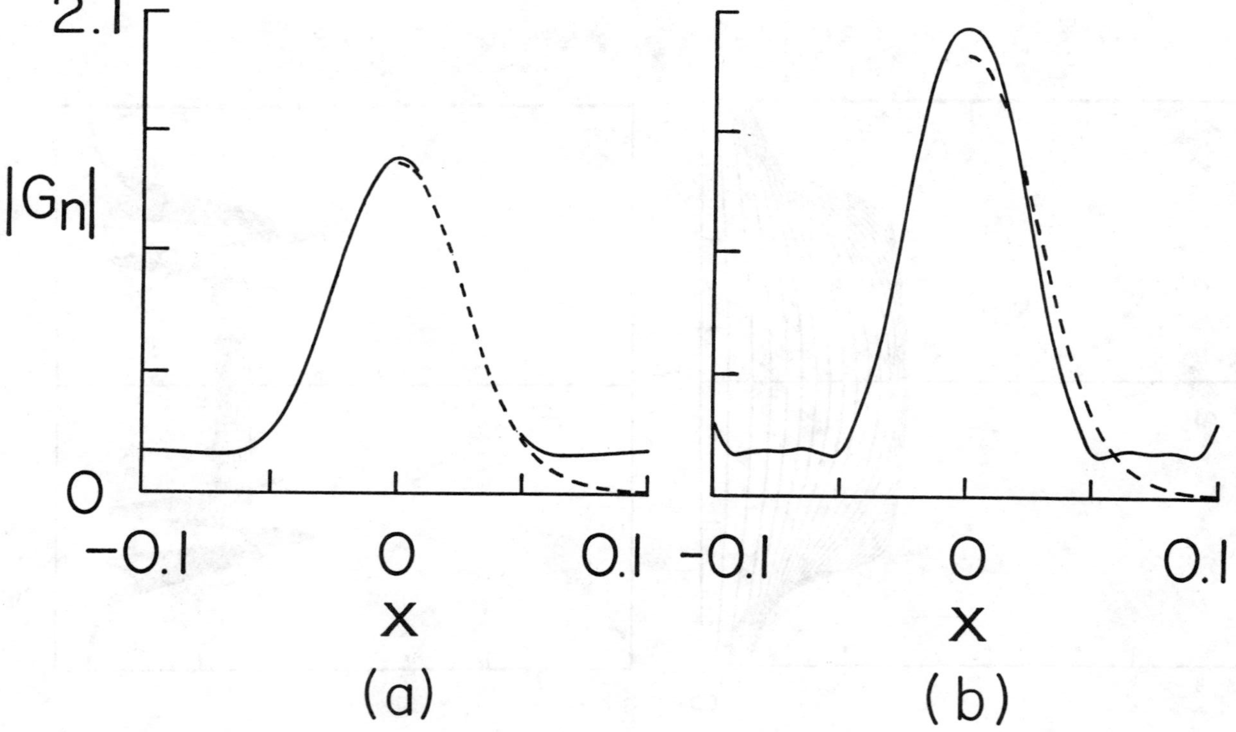

Figure 9. Transverse profiles of the solitary wave fixed points. The dashed curves are theoretical predictions. a) A fixed point which contains one solitary wave. b) the central peak of a fixed point which contains seven solitary waves.

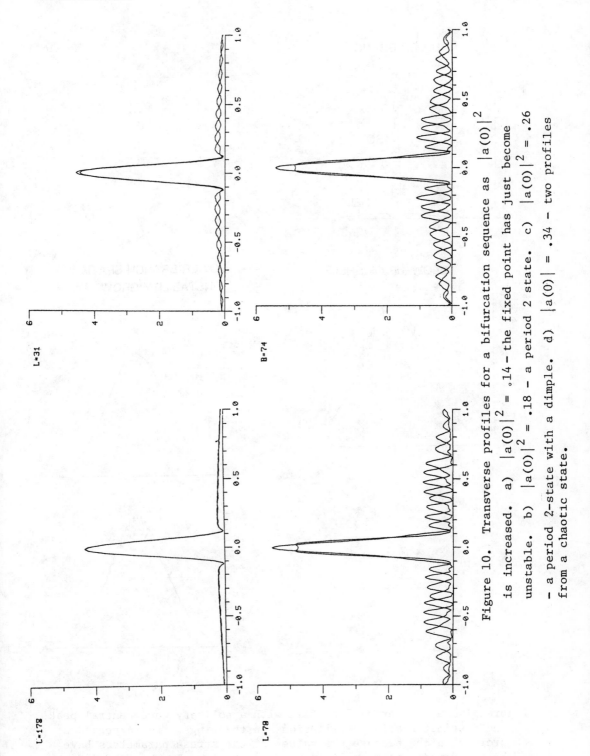

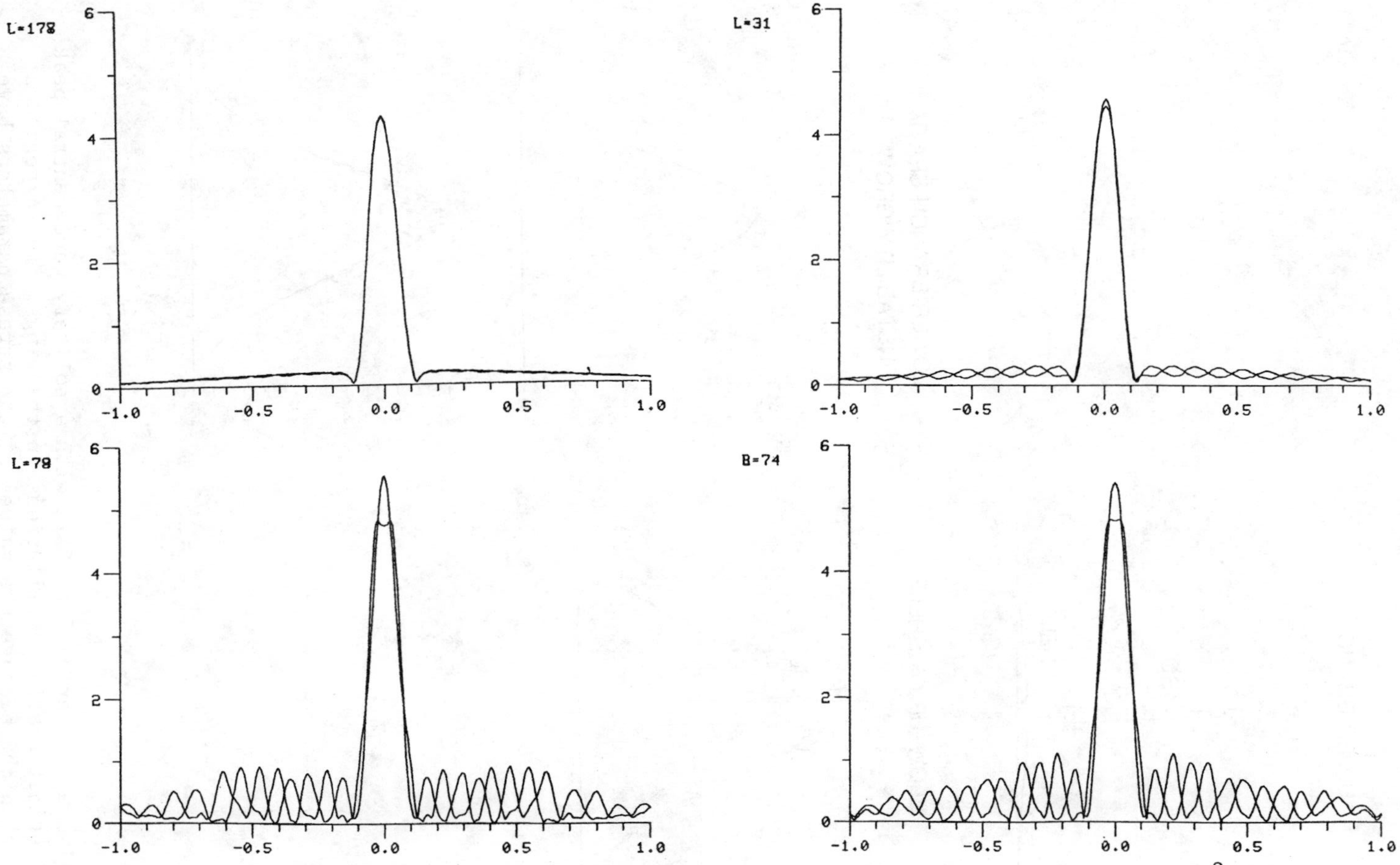

Figure 10. Transverse profiles for a bifurcation sequence as $|a(0)|^2$ is increased. a) $|a(0)|^2 = .14$ – the fixed point has just become unstable. b) $|a(0)|^2 = .18$ – a period 2 state. c) $|a(0)|^2 = .26$ – a period 2-state with a dimple. d) $|a(0)| = .34$ – two profiles from a chaotic state.

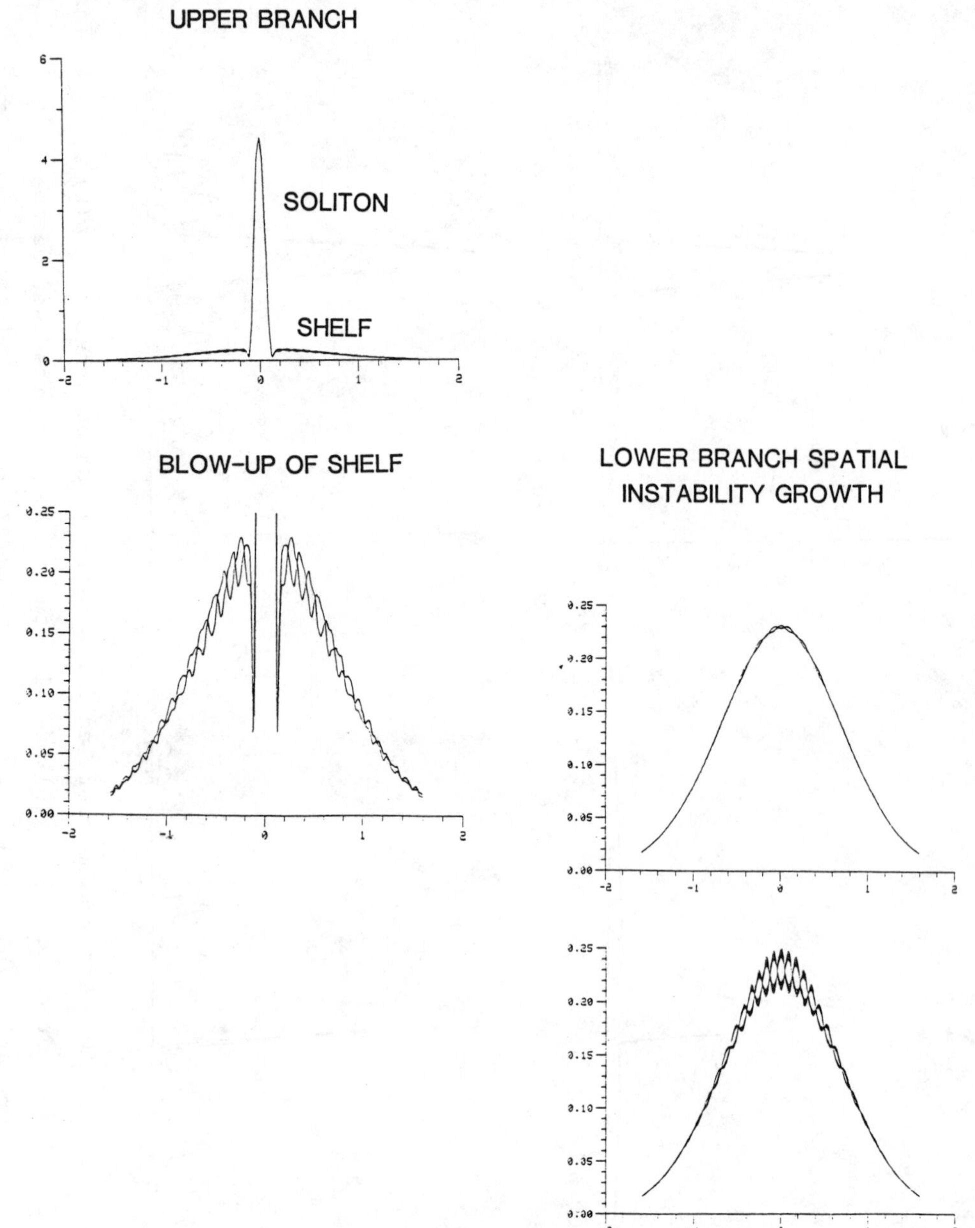

Figure 11. a) A period 2 profile with a solitary wave central peak and distinct spatial oscillations in the wings. b) A period 2 profile which, because the values of the stress parameters have been lowered, has no solitary wave central peak.

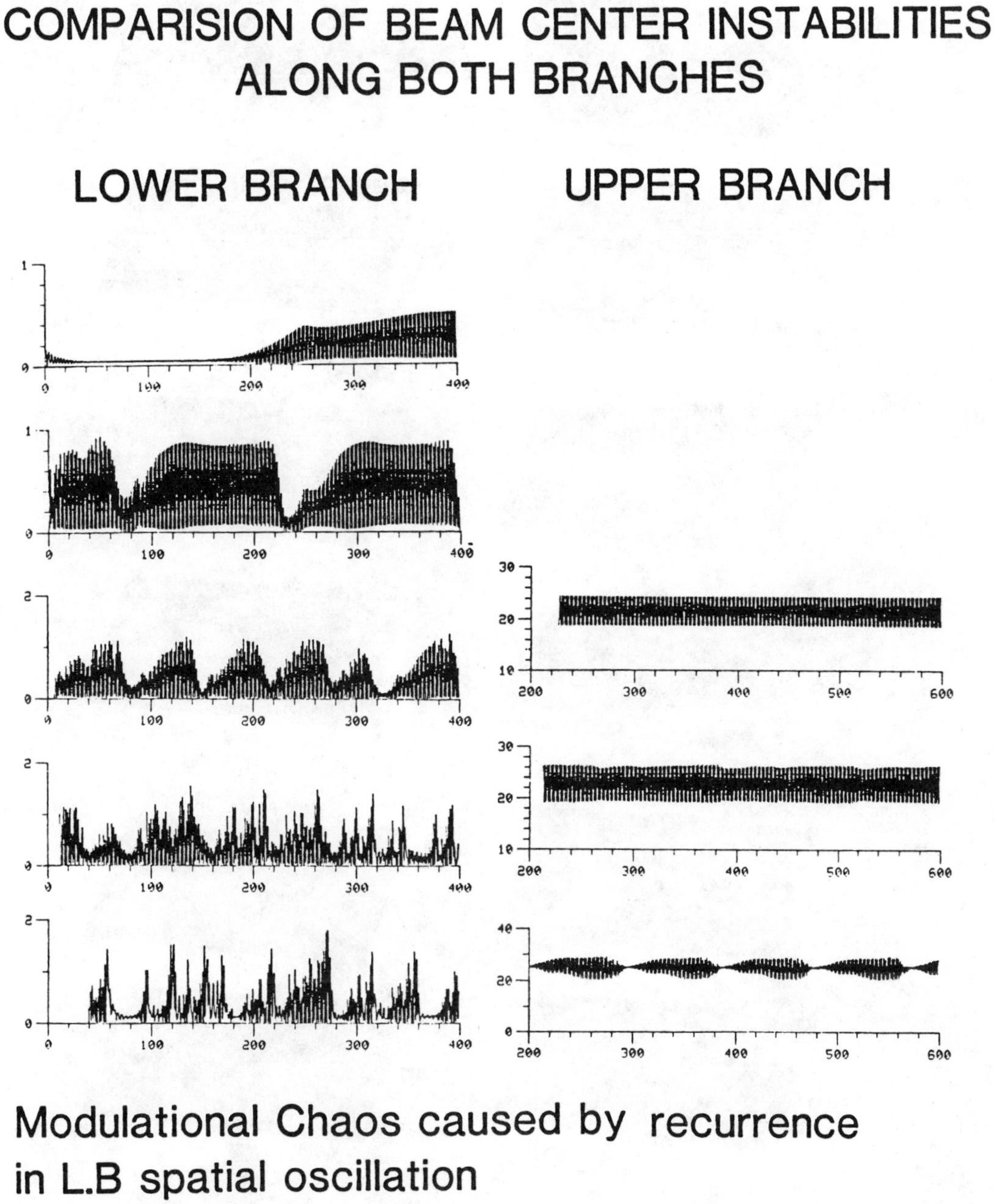

Figure 12. The peak amplitude as a function of n. Note the modulational chaos. The "upper branch" is data with a solitary wave, while the "lower branch" is data for an experiment with no solitary wave. After an initial transient, the figures show period 2 states and states with modulational chaos.

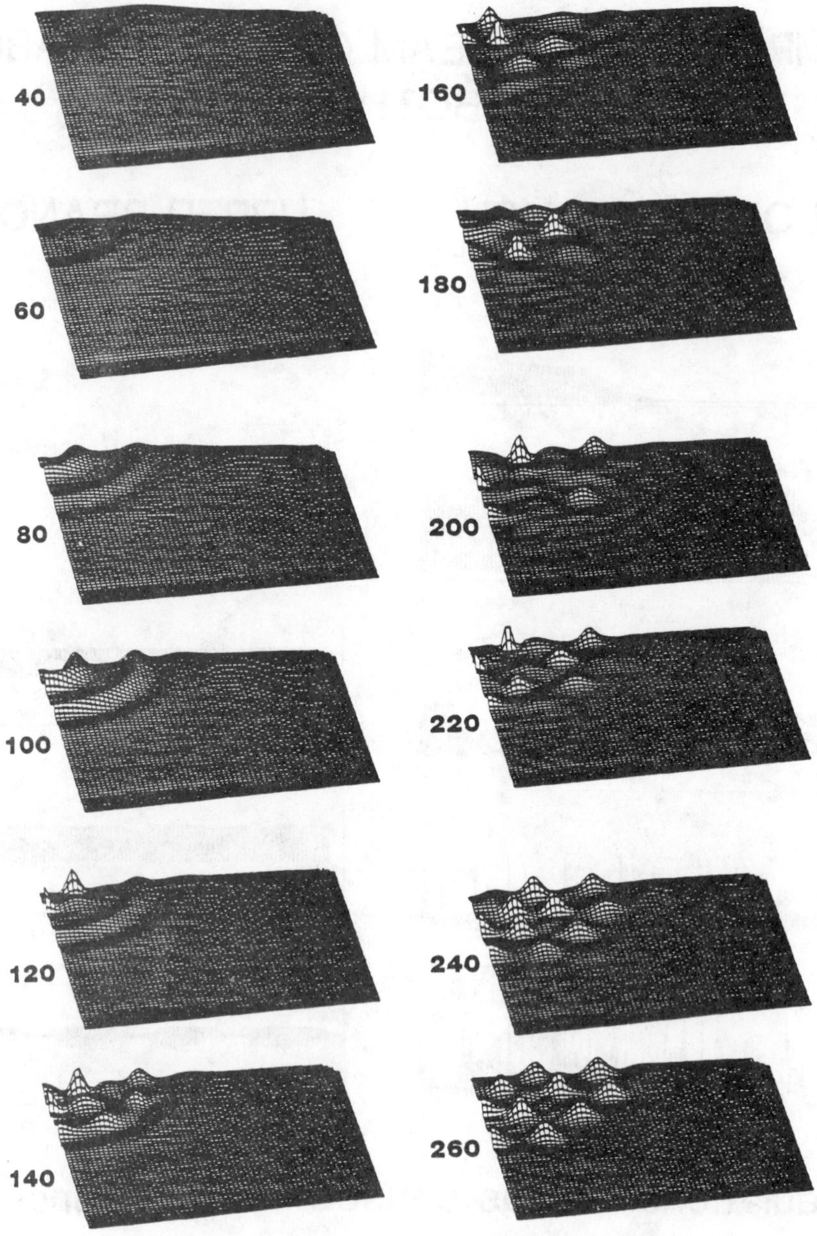

Figure 13. A two dimensional numerical experiment for the Kerr nonlinearity. The output $|G_n(x,y,z = \ell)|$ is depicted for passes $n = 40$ through $n = 260$.

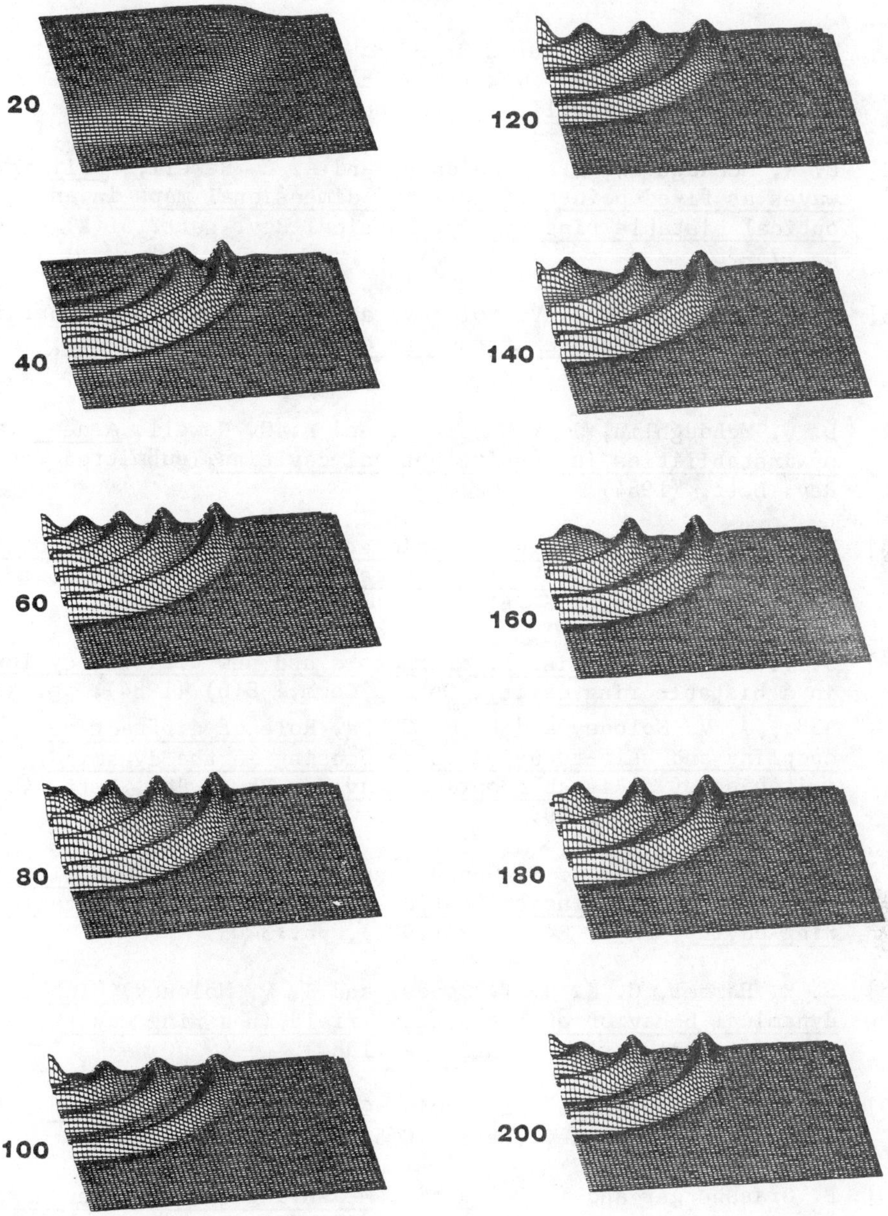

Figure 14. A two dimensional numerical experiment for saturable nonlinearity. $|G_n(x,y,z = \ell)|$ is depicted for passes n = 20 through n = 200.

REFERENCES

[1] C. M. Bowden, M. Ciftan, and H. R. Robl, Eds., Optical Bistability, I, Plenum, New York 1981; C. M. Bowden, Ed. Optical Bistability II, Plenum, New York, 1984.

[2] D. W. McLaughlin, J. V. Moloney, and A. C. Newell, Solitary waves as fixed points of infinite dimensional maps in an optical bistable ring cavity, Physical Rev. Lett. 51(2) (1983) pp. 75-78.

[3] D. W. McLaughlin, J. V. Moloney, and A. C. Newell, An infinite dimensional map whose attractors contain solitary waves, in preparation (1984).

[4] D. W. McLaughlin, J. V. Moloney, and A. C. Newell, A new class of instabilities in passive optical cavities, submitted Phys. Rev. Lett. (1984).

[5] K. Ikeda, Multiple valued stationary state and its instability of the transmitted light by a ring cavity system, Optics. Comm. 30(2) (1979), pp. 256-261.

[6] J. V. Moloney, Coexistent attractors and new periodic cycles in a bistable ring cavity, Optics Comm. 48(6) (1984), pp. 435-438; J. V. Moloney and H. M. Gibbs, Role of diffractive coupling and self-focusing on defocusing in the dynamical switching of a bistable optical cavity, Phys. Rev. Lett. 48(23) (1982), pp. 1607-1609.

[7] H. J. Carmichael, R. R. Snapp, and W. C. Schieve, Oscillatory instabilities leading to 'optical turbulence' in a bistable ring cavity, Phys. Rev. A 26(1982), pp. 3408.

[8] S. M. Hammel, C. K. R. T. Jones, and J. V. Moloney, Global dynamical behavior of the optical field in a ring cavity, preprint, University of Arizona (1984).

[9] P. Grassberger and I. Procaccia, Characterization of strange attractors, Phys. Rev. Lett. 50(5) (1983), pp. 346-349.

[10] P. Grassberger and I. Procaccia, Measuring the strangeness of attractors, Physica 9D(1983), pp. 189-708

[11] K. J. Blow and N. J. Doran, Global and local chaos in the pumped nonlinear Schrödinger eq., Phys. Rev. Lett. 52 (1984), p. 526.

[12] D. W. McLaughlin, J. V. Moloney, and A. C. Newell, in preparation 1984.

NOISY SWITCHES

FRANK MOSS*

Abstract. Switches are, perhaps, the most fundamental component of any machine --biological or man made. The simplest switch is a bistable system, capable of two states: "on" or "off". In this article, I explore, from both theoretical and experimental points of view, two very different types of switches, and their behavior when exposed to large amplitude, external interference or noise. The technological implications are fairly obvious: ranging from computer memories (vast arrays of switches) which must preserve existing, or memorize new, information in the presence of cataclysmic external noise to problems of pattern recognition and image enhancement out of a noisy background, where, at every video or radar pixel a "switch" must indicate the presence or absence of a signal, and must make the decision in the face of intense noise. The scientific implications are equally intriging: spanning descriptions of first order, nonequilibrium phase transitions (which are often driven, or even stabilized, by external noise) to questions of prebiotic, molecular evolution.

1. Introduction. The two switches are represented by generic equations. Here, I use the term "generic" to indicate firstly that a more fundamental mathematical description of the switches is not possible, and secondly to signify that, with suitable modifications, the equation can represent all such switches in a class. They are first order, nonlinear, stochastic, differential equations with either parametric (multiplicative) or additive noise. In the first the nonlinearity is cubic. Bistability is inherent in this equation. In the second, the nonlinearity is only quadratic, and the bistability is the result of the noise itself. This device cannot perform as a switch in the absence of external noise.

In all cases, I use a stationary Fokker-Planck analysis in the white noise approximation to model the behavior of the switches. I have tested the models experimentally by building an electronic

*University of Missouri at St. Louis
St. Louis, MO 63121

analog of each switch, driving it with a noise generator, and analyzing the output with a computer. The computer (a Nicolet 1080) is specially designed for such experiments. My collaborators and I have developed a variety of programs specifically tailored for stochastic analysis of analog, experimental data. We can measure, and suitably average, the stationary statistical densities of one and two dimensional systems, the time evolving, drifting and diffusing densities after some initial condition is set and released or while some parameter is being swept, and the auto- and cross- correlation functions of statistical variables of interest.

Our results and techniques constitute a fundamental method for the study of noisy switches and the characterization of their behavior.

2. The white noise, Fokker-Planck model. In this section I shall briefly review the mathematical basis for the switch models. In the case of multiplicative noise, we shall always be dealing with stochastic differential equations of the form

$$dX/dt = h(X) + \lambda_t\, g(X), \tag{1}$$

where X is a variable of interest (say the current through a tunnel diode, for example) and λ_t is a control parameter which is noisy, that is

$$\lambda_t = \langle\lambda\rangle + \sigma\xi_t. \tag{2}$$

Here $\langle\lambda\rangle$ is the mean value of the control parameter, and is the only part which is externally controllable. ($\langle\lambda\rangle$ could be the voltage across the tunnel diode, (that is, the setting on a potentiometer) or it could be the mean position of a spring switch lever being manipulated by a nervous person.) The random part of λ_t is $\sigma\xi_t$ which we have no control over. In this analysis, I assume $\sigma\xi_t$ to be Gaussian, white noise of variance σ. (By "white" I mean that the power spectrum is a constant for $\omega \to \infty$, or equivalently that $\langle\xi_t,\xi_{t'}\rangle = \delta(t - t')$ is its auto-correlation function.)

Eq. (1) can be rewritten as

$$dX/dt = f(X,\langle\lambda\rangle) + \sigma\xi_t\, g(X), \tag{3}$$

where f is the deterministic part and $\sigma\xi_t g$ is the stochastic part. X_t, of course, is now a stochastic variable. One could seek solutions in the form of time trajectories

$$X_t = \int f(X,\langle\lambda\rangle)dt + \sigma\int\xi_t g(X)dt \tag{4}$$

Eq. (4) is difficult to deal with mathematically, because we must choose some initial condition $X_t(0)$, and then each trajectory of finite length commencing at $X_t(0)$ (which are the only ones of physical interest) will be different. As we shall see below, Eq. (4) is easy to integrate electronically.

In order to circumvent these difficulties, instead of seeking the time trajectories X_t, one finds the probability densities $P(X_t)$ which measures the probability that a neighborhood X will be visited at a time t by the trajectory. The analysis is straight forward and has been well known for many years [1]. $P(X_t)$ is described by a Fokker-Planck equation

$$\partial_t P(X_t) = \partial_X[f(X,\langle\lambda\rangle) + (1/2)\sigma^2 g(X)\partial_X g(X)]P(X_t) + (1/2)\sigma^2\partial_{XX}g^2(X)P(X_t) \qquad (5)$$

This equation is impossible to solve for the time evolving $P(X_t)$ without approximation except for certain special cases [2]. (We show below, however, that $P(X_t)$ can be quite easily measured in experiments.) So one normally seeks a stationary solution $\partial_t P(X_t) = 0$ which determines the steady state density $P_s(X)$ toward which $P(X_t)$ will evolve after a long time. The stationary solutions of Eq. (5) are well known:

$$P_s(X) = [N/g^\nu(X)]\exp(2/\sigma^2)\int(f/g^2)dX, \qquad (6)$$

where N is a normalization constant ordinarily determined by requiring that $\int P_s dX = 1$. The exponent ν indicates which so called stochastic calculus was used to obtain the Fokker-Planck equation: $\nu = 1$ for the Stratonovich calculus and $\nu = 2$ for the Ito calculus. It has been shown on theoretical grounds that Nature behaves according to the Stratonovich formulation [3,4]. A recent experiment -- the first able to distinguish between the two alternatives -- has confirmed this result [5]. I have here written Eq. (5) in the Stratonovich form. Since, in this article, I shall deal only with the stationary densities in the models, we need only consult Eq. (6) with $\nu = 1$.

Having dispensed with these necessities, we may now discuss the first switch.

3. A noise stabilized switch. We discuss here a switch modeled by a generic cubic of general form and no particular symmetry:

$$dX/dt = -X^3 + P_t X^2 - QX + R, \qquad (7)$$

where the noisy control parameter is $P_t = \langle P\rangle + \sigma\xi_t$. This represents the simplest, general bistable system, and shows the usual features: a lower branch (say the "off" state), an upper branch ("on") with a region of overlap (bistability). It is characteristic that such systems are hysteretic. Figure 1 shows several characteristics. The curve marked $\sigma^2 = 0$ shows the steady state $(dX/dt = 0)$ behavior of Eq. (7) for zero noise intensity.

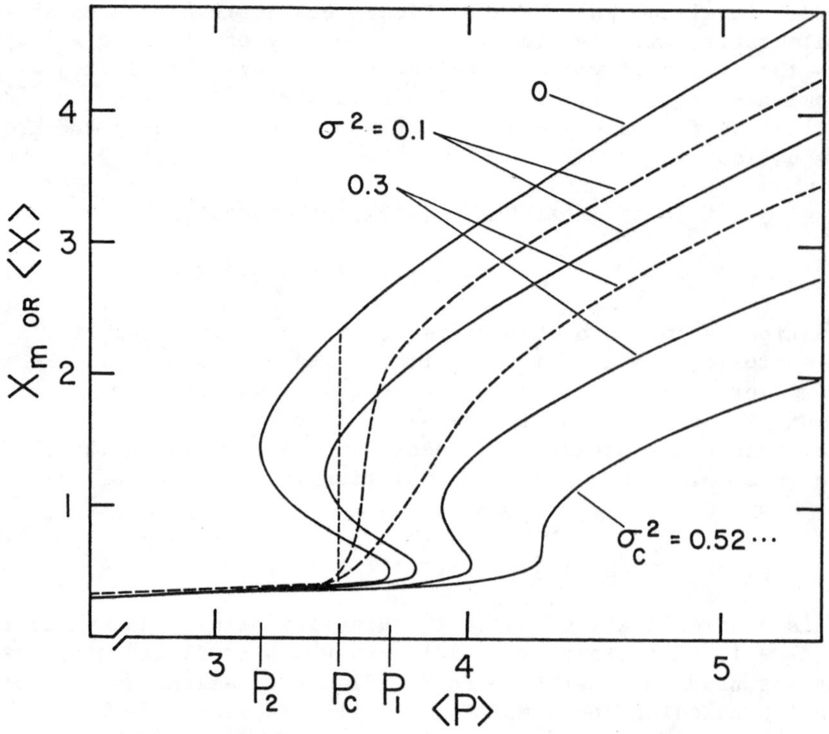

Figure 1. The steady states of X are shown by the curve $\sigma^2 = 0$. The other curves show the locations of the maxima X_m of the densities P_s for the indicated values of σ^2. The broken curves show $\langle X \rangle$.

Commencing from small values of $\langle P \rangle$, the system follows the lower branch until $\langle P \rangle = P_1$ whereupon it "switches" to the upper branch. The system point may then be moved along the upper branch at will, but in order to switch back to the lower branch $\langle P \rangle$ must be reduced to $\langle P \rangle = P_2 < P_1$, thus completing the hysteresis "loop". All good switches are quite obviously hysteretic.

The state of affairs as described above is, however, ideal in the extreme. We encounter trouble the moment we begin to enquire about the details of what actually happens at the switching point. (Is it a point after all? Is the system truly double valued at that point and nowhere else? -- excepting the other switching point, of course.) These questions cannot be answered in a practical way until we realise that the system is exposed to noise however small. The inherent bistability of the system is then revealed by P_s, which shows two modes so long as $P_2 < \langle P \rangle < P_1$. For this switch

$$f(X,\langle P\rangle) = -X^3 + \langle P\rangle X^2 - QX + R \tag{8a}$$

and

$$g(X) = X^2. \tag{8b}$$

Substituting these into Eq. (6) gives

$$P_s(X) = \frac{N}{X^2 + 2/\sigma^2} \exp\left[\frac{2}{\sigma^2}\left(-\frac{\langle P\rangle}{X} + \frac{Q}{2X^2} - \frac{R}{3X^3}\right)\right] \tag{9}$$

This result is shown in Fig. 2, where we have plotted a sequence of densities as <P> is increased from below the region of bistability to beyond it.

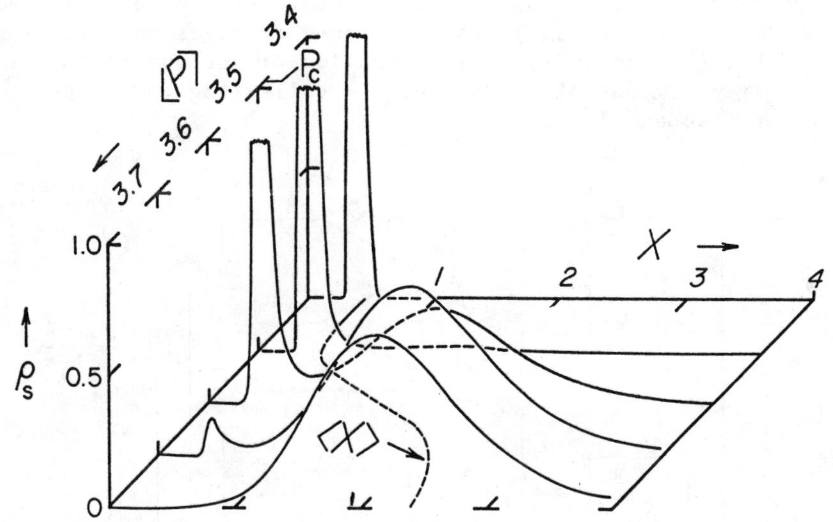

Figure 2. The evolution of the probability density as <P> is increased from a small value. This shows how a noisy switch actually achieves a transition from one state to another.

This shows how the probability of finding the switch on the lower branch shrinks with increasing <P> while the probability of finding it on the upper branch grows. This is how a noisy switch changes state. X_t, of course, is a fluctuating quantity. Its mean value is shown by the broken trace lying in the X, <P> plane.

Further insight can be obtained by locating the extrema of Eq. (9). For a given <P> within the bistable region, there are two maxima and a minimum located at X_m. We have previously shown [6] that the X_m in this model are determined by

$$(1 + \nu\sigma^2)X_m{}^3 - \langle P\rangle X_m{}^2 + QX_m - R = 0, \quad (10)$$

and in general by the solution of

$$\nu g'g = (2/\sigma^2)f. \quad (11)$$

The locations of these extrema are plotted in Fig. (1) shown by the three curves for $\sigma \neq 0$. The remarkable feature is that the effect of the noise is to push these curves <u>toward larger values of the control parameter</u>. <u>Noise</u>, therefore, <u>tends to stabilize the transition</u>, since larger values of <P> are required to achieve the same switching probability. In addition, the noise causes a reduction in the width of the bistable region, until at $\sigma_c{}^2 = [(Q/3)^3/R^2 - 1]/\nu$ the width $\to 0$ and the transition becomes second order.

I have designed an electronic version of Eq. (7), which is shown in Fig. 3. The system essentially assembles some voltages representing the right of Eq. (7), including multiplication of the quadratic term by a noisy voltage, then integrates them by collecting charge on a capacitor in a feedback loop.

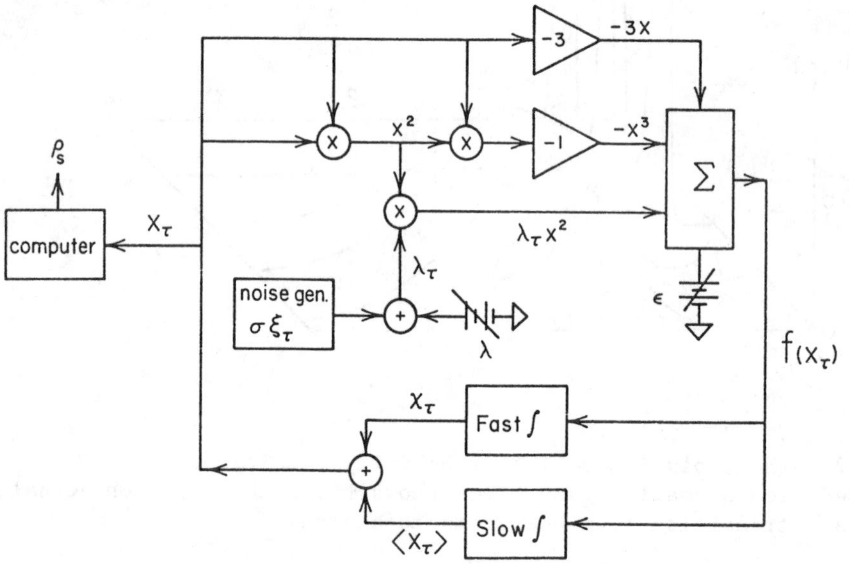

Fig. 3. An electronic system which imitates a generic switch represented by an asymmanetric cubic.

Measured results are shown in Fig. 4.

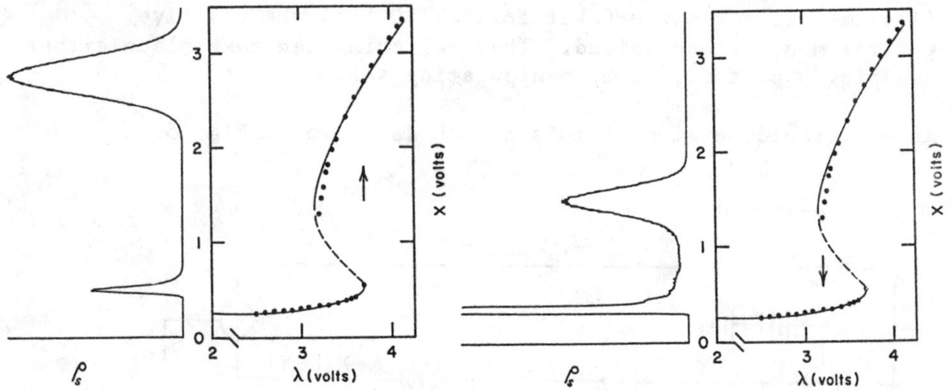

Fig. 4. These show switching transitions in the asymmetric cubic. On the left the transition is from lower to upper branch, and visa versa on the right. The solid S shaped curves are calculated from Eq. (7) with $dX/dt = 0$. The solid circles show the same result as measured on the electronic system. The measured, stationary densities are shown on the left.

A somewhat different version of Eq. (7) is obtained with $P = 0$ and using Q as a control parameter. For $R = 0$, this system is symmetric about the Q axis. For $R \neq 0$, the symmetry is broken. Theoretical studies [7-9] of this system have been applied to the important question of prebiotic molecular chirality evolution. In a recent experiment, the time evolution of the statistical density was measured for the first time, and the results so obtained established a firm foundation for the previous theoretical work [10].

4. <u>A switch which is not bistable without noise.</u> In this section I consider a dynamical system with a quadratic nonlinearity:

$$dX/dt = 1/2 - X + \lambda_t X(1-X), \tag{12}$$

where, as usual, λ_t is the noisy (multiplicative) control parameter. This system has been extensively studied, both theoretically [11,12] and experimentally [13]. Eqs. (6) and (12) with

$$f = 1/2 - X + \langle\lambda\rangle X(1-X) \tag{13a}$$

and

$$g = X(1 - X) \tag{13b}$$

result in the stationary density:

$$P_s = [X(1-X)]^{-\nu}\exp\{-[\sigma^2 X(1-X)]^{-1} - (2\langle\lambda\rangle/\sigma^2)\ln[(1-X)/X]\}. \tag{14}$$

This density exhibits a critical noise intensity σ_c. For $\sigma<\sigma_c$ P_s is monomodal, but for $\sigma>\sigma_c$ it is bimodal. In the bimodal regime, P_s is symmetric for $\langle\lambda\rangle = 0$ but for $\langle\lambda\rangle > 0$ (<0) the positive (negative) mode is emphasized. Thus switching can take place either by sweeping σ past σ_c, or by manipulating $\langle\lambda\rangle$.

An electronic version of this switch is shown in Fig. 5.

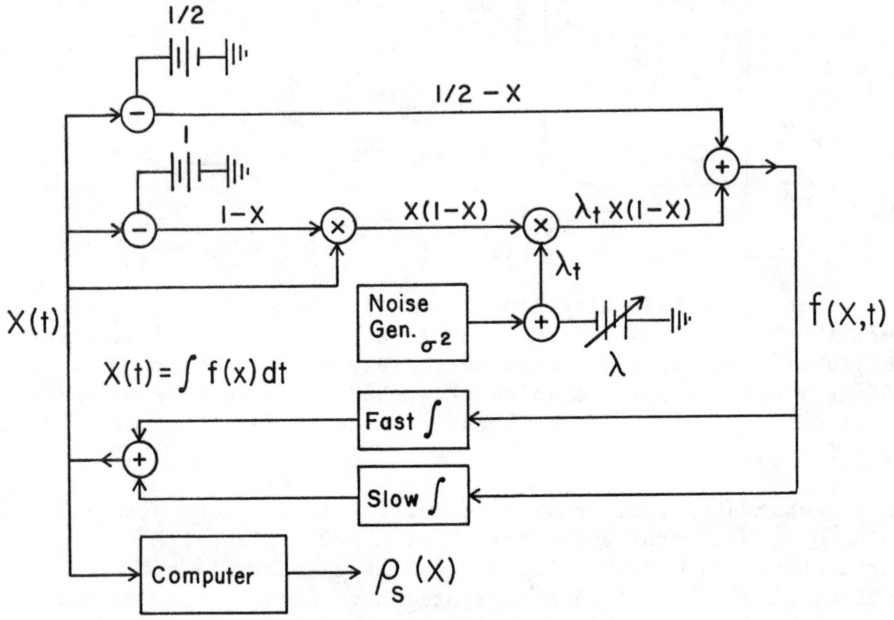

Figure 5. An electronic switch witn a quadratic nonlinearity. This is a noise induced switch.

Measurments of the stationary densities for $\langle\lambda\rangle = 0$ (symmetric case) are shown in Fig. 6, for various values of $\sigma>\sigma_c$ or $\sigma<\sigma_c = \sqrt{2}$. The dotted curves are theoretical predictions, and only those calculated from the Stratonovich calculus agree with the experimental results.

Figure 6. Measured densities (a) for $\sigma = \sigma_c$. The dotted curves are theoretical results. The bimodal dotted curve was calculated using the Ito calculus. It does not agree with the experimental result. (b) The monomodal density is for $\sigma = 1.5 < \sigma_c = \sqrt{2}$. The bimodal density is for $\sigma = 2.5 > \sigma_c$. This is noise induced switching for $\langle\lambda\rangle = 0$.

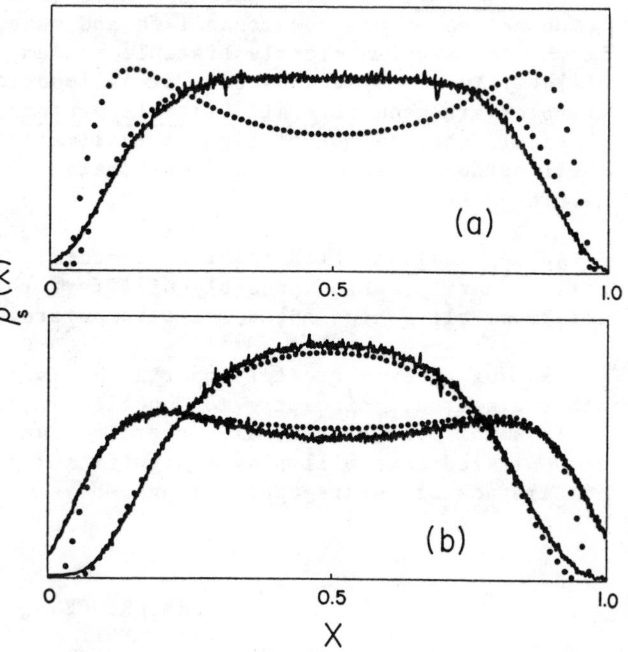

Fig. 7. shows the effect of changing $\langle\lambda\rangle$. Note that for $\langle\lambda\rangle \sim 1$, the positive mode is strongly selected. This is an example of control parameter induced switching.

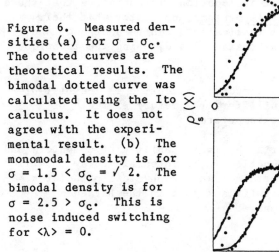

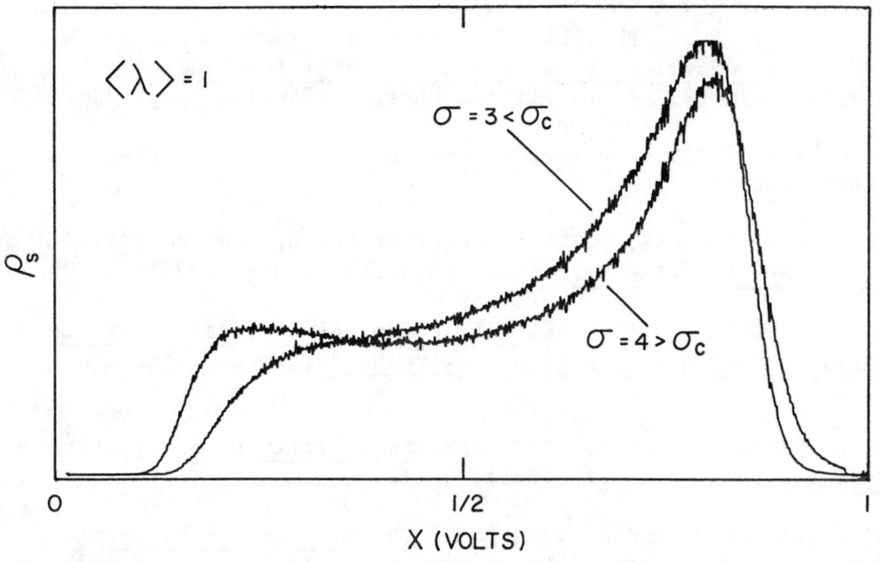

Figure 7. Control parameter induced switching.

5. Summary. I have discussed here two switches which are subject to noise, and shown how to analyze and measure their performance. The first one is an inherently bistable system, while in the second, bistability is noise induced. It is important to emphasize that these examples are generic. All bistable systems, no matter what their physical, chemical or biological realizations can be analysed, and their performances quantitatively characterised, using the techniques developed here.

An application which requires immediate attention is the bistable optical cavity, which possibly will form the memory elements of tomorrows ultra high speed macro-computers.

Perhaps of even greater importance, is the ability of computers and other electronic circuitry to function in an intensely noisy, hostile environment. It is virtually certain that the physical principles demonstrated here will play a significant role in the hardened electronics of future applications.

REFERENCES

1. N. G. VAN KAMPEN, Stochastic Processes in Physics and Chemistry, North Holland, Amsterdam, 1983, Chapt. VII.

2. N. G. VAN KAMPEN, A Soluble Model for Diffusion in a Bistable Potential, J. Stat. Phys. 17 (1977), p. 71.

3. B. J. WEST, A. R. BULSARA, K. LINDENBERG, V. SESHADRI and K. E. SHULER, Stochastic Processes with Non-additive Fluctuations, Physica, 97A (1979), p. 211 and Physica 97A (1979), p. 234.

4. N. G. VAN KAMPEN, Ito Versus Stratonovich, J. Stat. Phys. 24 (1981), p. 175.

5. J. SMYTHE, F. MOSS, P.V.E. McCLINTOCK and D. CLARKSON, Ito Versus Stratonovich Revisited, Phys. Lett. 97A (1983), p. 95.

6. G. V. WELLAND and F. MOSS, Enhancement of Critical Onsets in a Bistable System with Multiplicative Noise, Phys. Lett. 89A (1982), p. 273.

7. D. K. KONDEPUDI and I. PRIGOGINE, Sensitivity of Nonequilibrium Systems, Physica, 107A (1981), p. 1.

8. D. K. KONDEPUDI and G. W. NELSON, Chiral Symmetry Breaking in Nonequilibrium Systems, Phys. Rev. Lett. 50 (1983), p. 1023.

9. D. K. KONDEPUDI, in Fluctuations and Sensitivity in Nonequilibrium Systems, W. HORSTHEMKE and D. K. KONDEPUDI, eds., Springer-Verlag, Berlin, forthcoming; and Chiral Symmetry Breaking in Nonequilibrium Chemical Systems: Time Scales for Chiral Selection, submitted to Phys. Rev. Lett.

10. F. MOSS, D. K. KONDEPUDI and P. V. E. McCLINTOCK, Selectivity at a Noisy Bifurcation, submitted to Phys. Rev. Lett.

11. W. HORSTHEMKE and R. LEFEVER, Phase Transition Induced by External Noise, Phys. Lett., 64A (1977), p. 19.

12. R. LEFEVER and W. HORSTHEMKE, in Nonlinear Phenomena in Chemical Dynamics, C. Vidal and A. Pacault, eds., Springer-Verlag, Berlin, 1981, p. 120.

13. J. SMYTHE, F. MOSS and P.V.E. McCLINTOCK, Observation of a Noise-Induced Phase Transition with an Analog Simulator, Phys. Rev. Lett., 51 (1983), p. 1062.

GENERALIZED MULTISTABILITY AND CHAOS IN QUANTUM OPTICS

F. T. ARECCHI*

Abstract. A challenging problem in non-equilibrium statistical mechanics is that regarding the insurgence of ordered structures starting from a chaotic (maximum entropy) condition, in a system strongly perturbed at its boundary as a quantum optical system. For still higher perturbations, the ordered structures become more and more complex, until reaching deterministic chaos. Three experimental situations for CO_2 lasers (a laser with modulated losses, a ring laser with competition between forward and backward waves, and a laser with injected signal) are analyzed as examples of the onset of chaos in systems with a homogeneous gain line and with a particular time scale imposed by the values of the relaxation constants. I stress the coexistence of several basins of attraction (generalized multistability) and their coupling by external noise. This coupling induces a low frequency branch in the power spectrum. Comparison is made between the spectra of noise-induced jumps over independent attractors and that of deterministic diffusion within subregions of the same attractor. At the borderline between the two classes of phenomena a scaling law holds, relating the control parameter and the external noise in their effect on the mean escape time from a given stability region.

1. Introduction: order and chaos in quantum optics.

Quantum optics deals with lasers and laser-like phenomena. At an elementary level, they can be understood in terms of perturbation theory at lowest orders; that is, in terms of competition between stimulated and spontaneous emission processes.

If one tries to build a nonperturbative picture, one is struck by the complexity of the problem. There are two conceptual escape ways: on one hand to increase the size of the system to infinity

*Dept. of Physics - University of Firenze and Istituto Nazionale di Ottica - Firenze - Italy

while keeping the density finite and the temperature uniform, and look for an asymptotic (thermodynamic) solution; on the other hand to drastically simplify the boundary conditions making it possible to excite only one or a few radiation modes. This was the original idea of Schawlow and Townes, when extending the Maser principle to optical frequencies by use of a Fabry-Perot cavity. This is also done in other classes of nonlinear field problems such as hydrodynamical instabilities, where one works with small "aspect ratios", that is, with cells of comparable sizes in all three dimensions, in order to excite few Fourier components of velocity and thus deal with a finite number of coupled equations. In quantum optics this procedure leads to a set of quantum equations still unsoluble. A further approximation is the so-called "semi-classical" one, leading to the Maxwell-Bloch equations. Starting in the early sixties, introduction of photon statistics methods has made possible the study of fluctuations and coherence in lasers: how and why 10^{20} atoms or molecules, rather than radiating e.m. field in a chaotic fashion, decide to "cooperate" to a single coherent field mode; then, for still higher excitation, how and why they organize in a complex pattern (many modes), each per-se highly coherent but with little correlations with one another. The lack of large scale correlations is shown to correspond to the onset of deterministic chaos.

It is worthwhile to stress the role of nonlinearities as well as that of the phase space topology (number of relevant dynamical variables) for the onset of order and chaos. A detailed experimental analysis of the onset of order in a pumped system was given in my 1965-67 investigations on the passage from incoherent to coherent light in a laser [1]. Fig 1 shows the photon statistics (P.S.) for a

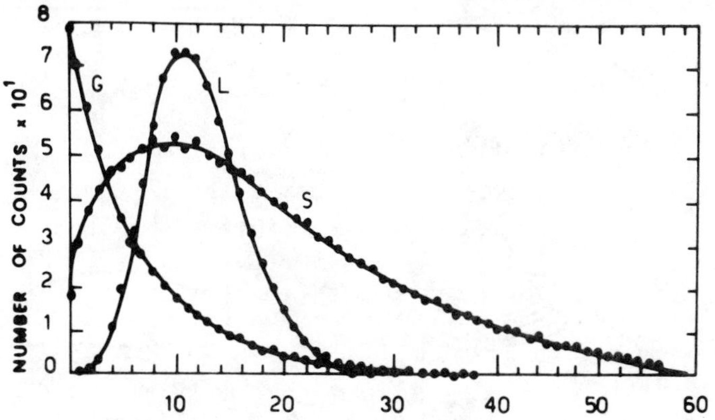

Figure 1.Photocount distributions of three radiations fields. L, Laser field; G, Gaussian field; S, linear superposition of L and G onto the same space mode.

radiation field below and above the threshold point (that is, the point where the gain provided by the stimulated emission processes of excited atoms compensates for the losses due to the escape of radiation from the laser volume). The G curve is fitted by a Bose-Einstein distribution describing the fluctuations of the photon number in a black-body around the average value $\langle n \rangle$ given by Planck's formula, the L curve is fitted by a Poisson distribution. The two distributions correspond to fields with the same color, direction and intensity, so that there is no classical optics measurement which could discriminate between them. Yet the P.S. mesurement shows a dramatic difference. The reason why is given in Fig. 2. For a uniform field, since photons are Bose particles with zero mass, and hence delocalized, the associated photon detection processes at different points in space-time have no correlations. Therefore, in a volume filled with a coherent, or ordered, field (that is, a field with δ-like statistics as in Fig. 2a), the associated probability p(n) of detecting photons at a given point over a time T is a Poisson distribution.

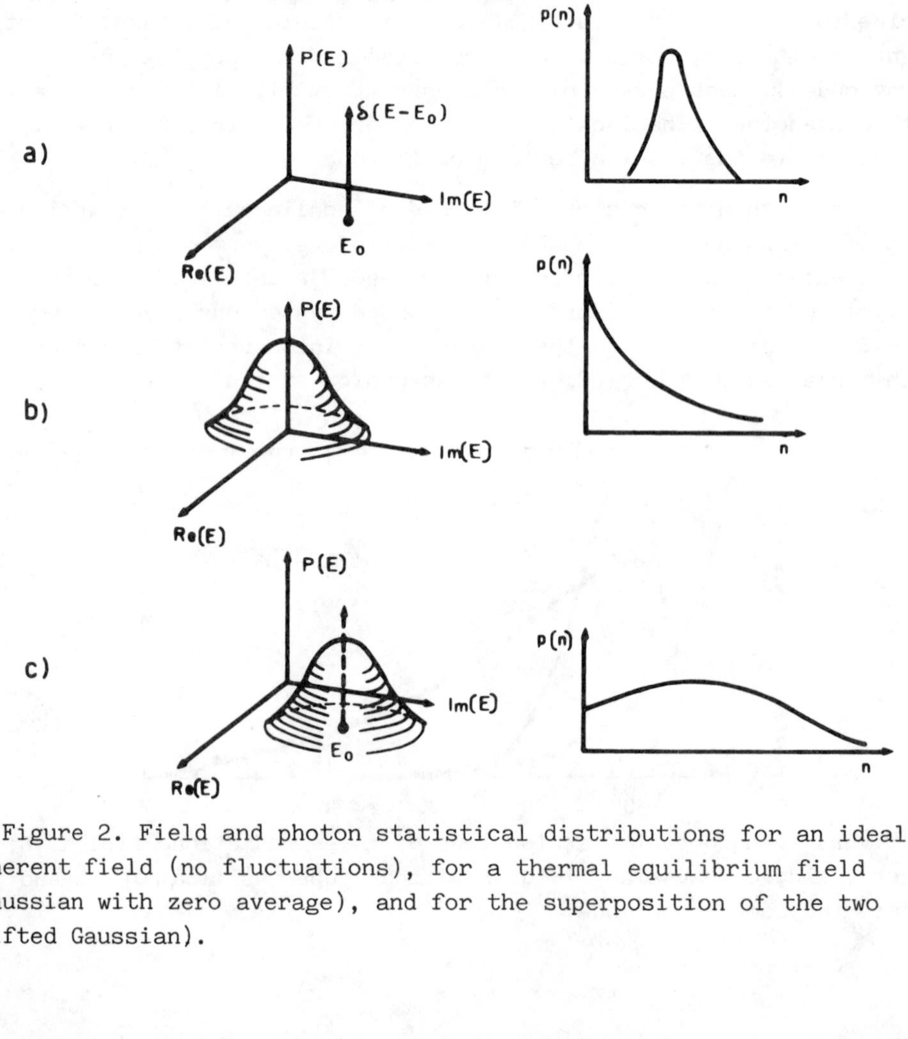

Figure 2. Field and photon statistical distributions for an ideal coherent field (no fluctuations), for a thermal equilibrium field (Gaussian with zero average), and for the superposition of the two (shifted Gaussian).

If now the field has a zero-average Gaussian distribution, as for a thermal equilibrium or maximum entropy situation, weighting each probability element with the detection statistics one obtains the Bose-Einstein distribution (Fig. 2b). Between these two limiting cases of full order and maximum chaos, one can trace a continuous manifold of intermediate cases (Fig. 2c). This smooth behavior is analogous to a second order phase transition , as when a thermodynamic system undergoes a continuous change of state around a critical temperature. The explanation implies the essential role of nonlinearities. The elementary description of a laser in terms of Einstein stimulated emission processes compensating for losses is not sufficient. Indeed, this would just provide a linear polarization $P = \chi E$, and a quadratic free energy

$$F(E) = - \vec{P}\cdot\vec{E} = - \chi E^2. \tag{1}$$

In a thermodynamic system open with respect to the field variable E, this has a statistical distribution given by

$$W(E) = Ne^{-F(E)/kT}, \tag{2}$$

(N = normalization constant).
Similarly a nonequilibrium system with a dynamical variable E, driven by a nonlinear force f(E) and by stochastic noise with short correlation time and correlation amplitude D, has a stationary distribution [2,3].

$$W(E) = Ne^{-\int f(E)dE/D}. \tag{3}$$

In the absorbing case (force of the field proportional to the polarization $P = -\alpha E$; $f(E)dE = -\frac{1}{2}\alpha E^2$) by Eq. (3) the field has a Gaussian stationary distribution, as it should be expected from thermodynamics. In the linear emitting case the distribution is undefined ($P = \alpha E$; $e^{\alpha E^2}$ is not normalizable).

But an atom is still exposed to photons after emission. The lowest correction is cubic in the field (Fig. 3)

$$P = \alpha E - \beta E^3$$

and it is sufficient to describe the passage from Guassian chaos to a narrow distribution around nonzero fields

$$\pm E_o = \pm \sqrt{\alpha/\beta} \tag{4}$$

The spontaneous symmetry breaking does not assign the phase of E_o. We have two equivalent states 180° apart. To lift the degeneracy we must apply an external field γ , which assigns a reference phase. The two states are no longer equivalent (optical bistability)(Fig. 3d).

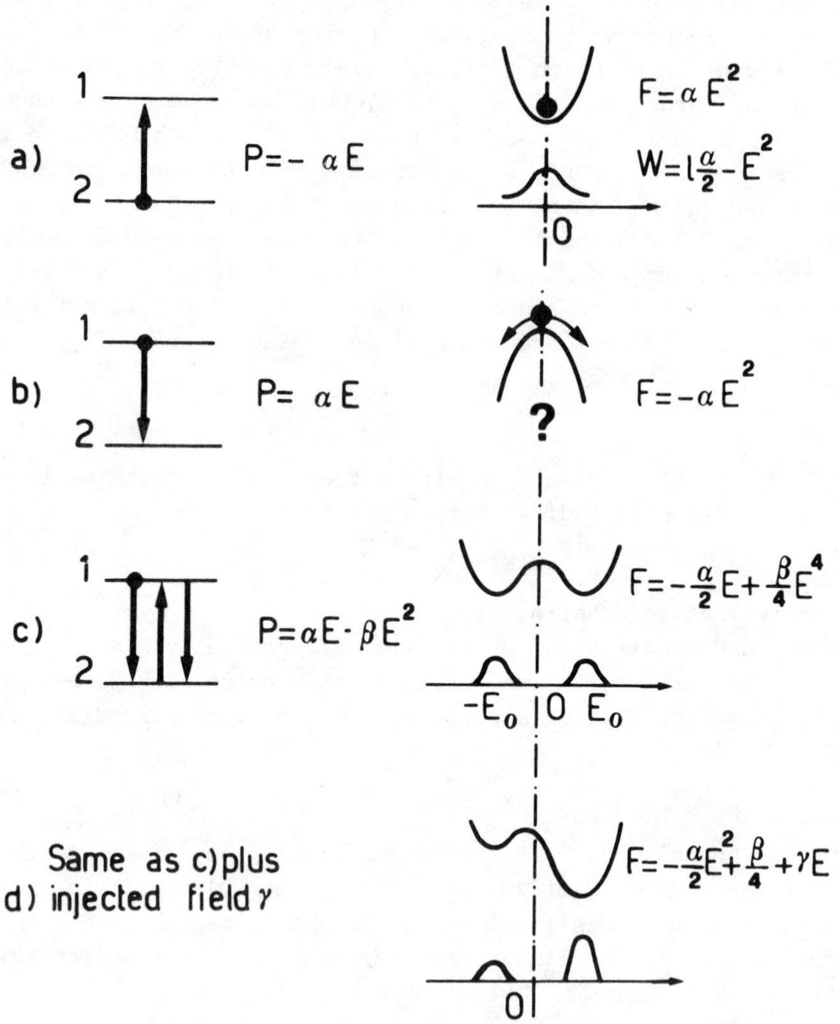

Figure 3. a) Absorbing atom, linear polarization; parabolic pseudo-potential; Gaussian probability.
b) Emitting atom, linear polarization; parabolic pseudo-potential; undefined probability.
c) Emitting atom, cubic polarization; quadratic pseudo-potential; probability peaks around values $\pm E_o$.
d) As c) but with a symmetry-breaking odd term in the potential, arising from an external field.

As shown by the locus of stable points, at threshold the thermodynamic branch (Fig. 3a) becomes unstable and the "coherent" branches appear (Fig. 3c). In mathematical terms, this is a bifurcation. Still higher order bifurcations could appear, making the ordered branch unstable, and leading to new "orders".

Before discussing this multiple sequence of bifurcations it is important to decide how many degrees of freedom we have to deal with. In physics we deal in general with nonlinear equations for a field $\vec{q}(x,t)$

$$\frac{\partial \vec{q}}{\partial t} = f\left(\vec{q}\,, \frac{\partial}{\partial x}\vec{q}\right). \tag{5}$$

Such are Navier-Stokes and Fourier equations for a velocity field coupled to a temperature field in a convective fluid instability. For a rectangular cell of small aspect ratio (ratio of two lateral sizes with respect to the fluid height) and for a temperature difference ΔT between lower and upper plate near the onset of the instability, eqs. (5) reduce to three coupled equations for a velocity mode and two temperature modes [4]. In suitable units, these three equations are[5].

$$\begin{aligned} \dot{x} &= -\sigma(x-y) \\ \dot{y} &= -y + r\,x - xz \\ \dot{z} &= -(8/3)z + xy \end{aligned} \tag{6}$$

With $\sigma = 10$ and $r = 28$ the solution is chaotic.

Similarly, if we couple Maxwell equations with Schrödinger equations for N atoms confined in a cavity, and we expand the field in cavity modes, keeping only the first mode E which goes unstable, this is coupled with the collective variables P and Δ describing the atomic polarization and population inversion as follows [2,3]

$$\begin{aligned} \dot{E} &= k\,E + g\,P \\ \dot{P} &= -\gamma_{\perp}P + g\,E\Delta \\ \dot{\Delta} &= -\gamma_{\parallel}(\Delta - \bar{\Delta}) - 2g\,(P^*E + PE^*) \end{aligned} \tag{7}$$

where k, $\gamma_{\perp}$, $\gamma_{\parallel}$ are the loss rates for field, polarization and population respectively, g is a coupling constant and $\bar{\Delta}$ is the population inversion which would be established by the pump mechanism in the atomic medium, in the absence of coupling. While the first of eq. (7) comes from Maxwell eqs., the two others imply the reduction of each atom to the levels which are resonantly coupled with the field, that is, a description of each atom in a isospin space of spin 1/2. The last two eqs. are Bloch eqs. which describe the spin precession. Therefore, eqs. (7) are called Maxwell-Bloch equations.

Eqs. (6) and (7) are phenomenological. The presence of loss rates means that three relevant degrees of freedom are in contact with a "sea" of other degrees of freedom. In principle, they could be deduced from microscopic equations by statistical reduction techniques. The fluctuation-dissipation theorem would impose the addition of stochastic forces.

However, we show that for $N \geq 3$ degrees of freedom, deterministic chaos may be reached in nonlinear eqs. (6) or (7) without consideration of stochastic forces. These latter ones would modify some details of the phenomena, without relevant changes in the qualitative picture. In a dissipative system there is a contraction of the phase space volume. If at time $t = 0$ the ensemble of initial conditions is confined in a hypersphere of radius ϵ, (initial volume $V_0 = \epsilon^N$) at time t the volume will be (referring to the principal axes)

$$V(t) = \epsilon^N e^{\sum_1^N \lambda_i t}, \tag{8}$$

where the growth rates λ_i are the Lyapunov exponents. The contraction requirement means

$$\sum_1^N \lambda_i < 0 . \tag{9}$$

Now, if we start from a single point at $t = 0$ (well determined initial condition) a single trajectory emerges, and on it we have obviously $\lambda = 0$. For $N = 1$, eq. (9) imposes $\lambda_1 < 0$ hence there is no trajectory. Indeed, the asymptotic volume can only be 0-dimensional, that is, a single point, hence we have only a stationary solution. For $N = 2$, if $\lambda_1 < 0$, it may be that $\lambda_2 < 0$ (no trajectory) but also $\lambda_2 = 0$, that is, the final volume is 1-dimensional (limit cycle) and we have a periodic oscillation. For $N = 3$, besides fixed points ($\lambda_1, \lambda_2, \lambda_3$ all negative) and limit cycles (λ_1, λ_2 negative, $\lambda_3 = 0$) we may have $\lambda_1 < 0$, $\lambda_2 = 0$, $\lambda_3 > 0$, with $\lambda_3 < |\lambda_1|$ to satisfy eq.(9). This means that in direction 3 we have a stretching from ϵ to $\epsilon e^{\lambda_3 t}$. Even if two initial conditions are very near, the representative points after a long time will be largely distant. This sensitive dependence on the initial conditions is the indicator of deterministic chaos.

The asymptotic phase space locus (after a long transient) for Lorenz eqs (6) is well known. That locus attracts all neighboring initial conditions because of the compression of the phase volume peculiar of dissipative systems. It is then an attractor, as the fixed point for $N = 1$ or the limit cycle for $N = 2$. But nearby points at a given time must diverge after a long interval, because of $\lambda_3 > 0$. Hence the attractor will never close on itself and it is called strange. The unpredictable behavior of paths started from initial conditions specified with an arbitrary (but finite) precision is a fundamental obstacle to long-term nonprobabilistic forecasting. This is why computers can produce only a sufficiently short realization of the path of a dynamic system with one (or more) $\lambda > 0$ (of course, the machine may go on computing, but, for large t, the path is no longer related to the initial conditions).

As for the physical realizability of deterministic chaos, eqs. (7) indicate that lasers are candidates for large varieties of situations as one changes the atomic species $(g, \gamma_\perp, \gamma_\parallel)$ the pump rate ($\bar{\Delta}$) or the losses of the e.m. cavity (k) [6]. On the other hand it is well known that commercial laser sources are good examples of stable dynamical systems. The main reason is that of time scales. As shown by the coefficients of eqs. (6), a strange attractor is obtained when the damping rates of x, y, z, are comparable.

On the contrary for noble gas lasers (Ne ,Ar) the atomic damping rates are much faster ($\gamma_\perp \sim \gamma_\parallel \sim 10^8 - 10^9 s^{-1}$) than the field loss rates ($K \sim 10^6 - 10^7\ s^{-1}$). Hence the second and third of eqs. (7) can be solved at equilibrium ($\dot{P} = \dot{\Delta} = 0$) with respect to the rather slow variations of E and substitution into the first yields a single non-linear dissipative equation $\dot{E} = f(E)$ which allows only for a fixed point. The procedure is called adiabatic elimination [2] of the fast variables, which are slaved by the slow variable. This latter one can be considered as the only relevant dynamical variable (order parameter), as it was implicit in the heuristic considerations of Fig. 3. We call these lasers Class A.

In some other systems (Class B, as e.g. ruby, Nd and CO_2 lasers) the population decays slowly, so that the dynamics is described by two coupled rate equations which are still insufficient to permit dynamical chaos. We call Class C systems those ones for which the three decay rates for the polarization, population and field are of the same order of magnitude. Thus far, no instabilities have been reported for simple Class C systems.

In order to increase by at least 1 the number of the degrees of freedom of Class B systems we must either

i) make the systems non autonomous, by a time dependent parameter, or

ii) increase the number of lasing modes, or

iii) increase the number of independent gain packets, that is, using an inhomogeneous line.

Point i) was shown in Ref. 7. Point ii) has been shown for two counter-rotating modes in a ring cavity [8,9] . Point iii) was treated in experimental and theoretical papers [10,11], without however a simple analysis in terms of a closed set of equations. In such a case, due to the large number of available packets, distinction in Class A and B is immaterial. On the contrary, as pointed out in Ref. 7, the beauty of a single mode laser with a homogeneous line is that the relevant instabilities can be modelled with the small number of equations (7) resembling the Lorenz model.

Let us first describe case i). We have chosen [7] a CO_2 laser system because the relaxation time $1/\gamma_{\parallel}$ of the population inversion is much larger ($1/\gamma_{\parallel} = 0.4$ ms) than the memory time $1/\gamma_{\perp}$ of the induced dipole ($1/\gamma_{\perp} = 10^{-8}$s), thus reducing the single-mode dynamics to the coupling between two degrees of freedom, namely photon population n and molecular population inversion Δ. Introducing within the cavity a time-dependent perturbation by an electro-optical molulator driven by a sinusoisal frequency, we have a non-autonomous differential system in n and Δ amounting to the crucial three degrees of freedom necessary to obtain chaos. As shown later, the relevant range of modulation frequencies is correlated to the value $\sqrt{K\gamma_{\parallel}}$; hence the choice of CO_2 laser has put the working frequency range in the easily accessible 50-130kHz region.

The coupled field-molecules equations for a single-mode laser lead, after adiabatic elimination of the polarization,reduce to the following rate equations:

$$\dot{\Delta} = R - Gn\Delta - \gamma_{\parallel}\Delta \quad , \quad \dot{n} = 2Gn\Delta - K(t)n, \qquad (10)$$

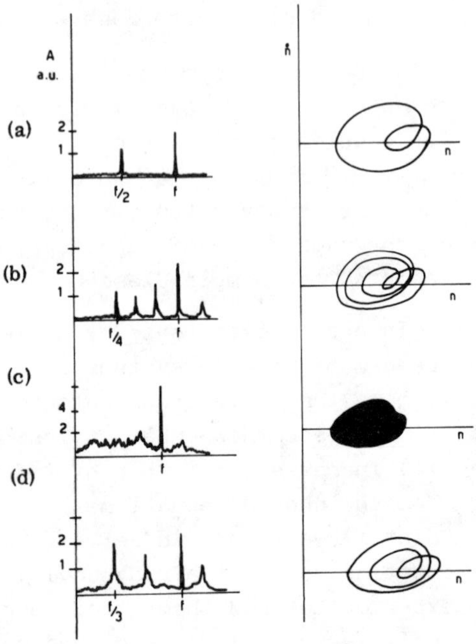

Figure 4. Experimental phase-space portraits ($\dot{n}$-n) (right side) and the corresponding frequency spectra (left side) for different modulation frequencies f. (a) f = 62.75 kHz. Period-two limit cycle and corresponding f/2 subharmonic. (b) f = 68.80 kHz. Period-four limit cycle and f/4 subharmonic. (c) f = 64.00 kHz. The phase-space portrait shows a strange attractor (the oscilloscope spot could not resolve single windings). The power spectrum is a quasicontinuous one with a small peak at the modulation frequency (see the scale change with respect to previous figures). (d) f = 64.13 kHz. Period-three limit cycle and f/3 subharmonic.

where

$$G = \omega\mu^2/\hbar\epsilon_0\gamma_\perp V = 0.25 \times 10^{-4}\,s^{-1} \qquad (11)$$

is the coupling constant (SI units) including the frequency ω and the dipole matrix element μ of the transition and the collisional broadening rate $\gamma_\perp$; R is the pump rate; V, the cavity volume, and

$$K(t) = K_1(1+m\cos\Omega t) \qquad (12)$$

is the cavity damping rate, modulated by the inserted electro-optical device. A linear pertubation analysis around the steady-state values $\bar{\Delta}$, $\bar{n}$, yields a linearized frequency value $\Omega_0/2\pi \simeq 43$kHz for the following typical parameters: $K_1 = 3 \times 10^7\,s^{-1}$ (corresponding to a cavity length of 2 m with losses of 20% per pass), and pumping rate R such that $\bar{n} \simeq \gamma_\parallel/G \simeq 10^8$ (corresponding to a dc power output of 50 μW). We select the two modulation parameters m and Ω as follows. We set K_1 consistently below the maximum damping rate $K_0 = GR/\gamma_\parallel$ compatible with the fixed pump rate R. It is easily shown that the linearized eigenfrequency is

$$\Omega_0 \simeq (\gamma_\parallel K_1)^{1/2} \qquad (13)$$

provided we choose $K_1 \simeq K_0/2$ and a modulation depth m sufficiently small to have

$$(1/K_1)(dK/dt) = \Omega m < \gamma_\parallel . \qquad (14)$$

The driving frequency $f = \Omega/2\pi$ is chosen to vary in the region from $\Omega_0/2\pi$ on, that is, from 40 to 150 kHz. As a consequence, $m < \gamma_\parallel/\Omega \simeq 10^{-2}$. We have explored modulation values between 1 % and 5%. The range m = 1% - 5% does not display m dependence; therefore we limit ourselves to giving experimental results at m = 1% for various Ω values.

In Fig. 4 we show experimental data in a narrow region between 62.7 and 64.25 kHz where various bifurcations occur. This region is limited above and below by wide intervals with stable single-period limit cycles. Fig. 4(a) shows the f/2 bifurcation at f =62.7 kHz, Fig. 4(b) the f/4 case for f = 63.8 kHz; Fig. 4(c) shows the strange attractor and a broad-band spectrum for f = 64.0 kHz; and Fig. 4(d) shows the f/3 case for f = 64.2 kHz.

As for case ii), we show in Fig. 5 the experimental set up allowing for independent detection of forward and backward waves in a ring cavity [8]. The coupling of the two waves via the nonlinear gain medium gives rise to competition, as shown in the phase diagram of Fig. 6 (p and I being the pressure and current of the gas discharge in the gain cell: increasing p and I means roughly to increase the density

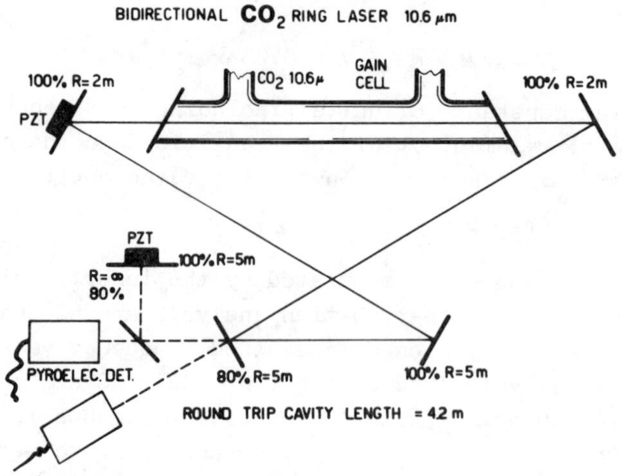

Figure 5. Bidirectional CO_2 ring laser system with external mirror.

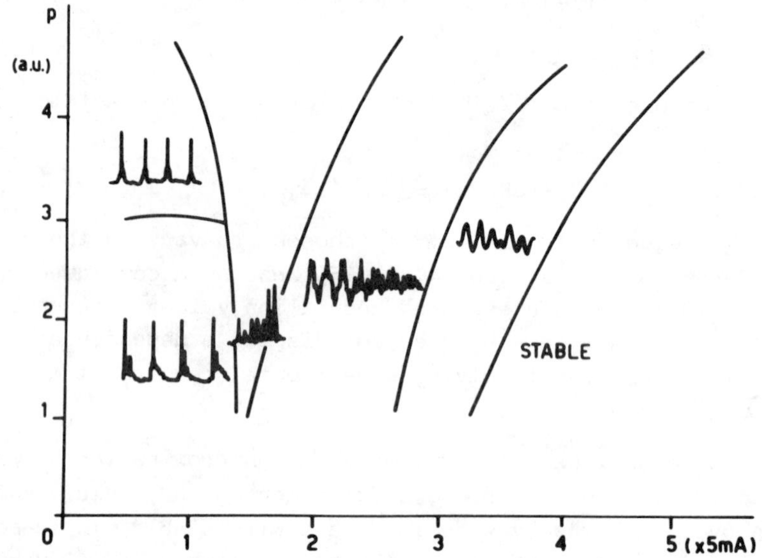

Figure 6. Phase space diagram for CO_2 ring laser; p and I are gas pressure and discharge current.

of excited atoms and hence the gain). Fig. 6 shows 5 different regions for increasing current. At low currents we have a regular spiking on both directions with a spike repetition given by $\gamma_{\parallel}^{-1}$. The spikes are synchronous for both fields. As we reduce the pressure, each spike develops a damped oscillation with a frequency $\Omega_0 \approx (k\gamma_{\parallel})^{1/2}$ as from eq. (13). This oscillation is 180° out of phase for the two fields, showing a coupling of one mode into the other. Increasing the current, we have an irregular transition region followed by a chaotic oscillation where the two modes alternate in a bistable fashion. This chaotic oscillation has a low frequency $\gamma_{\parallel}$ plus a superposed high frequency $(\gamma_{\parallel} k)^{1/2}$ which is evident in the third region but disappears in the fourth region. Measurement of the Grassberger-Procaccia correlation exponent shows a fractal dimension increasing from 2.5 to 5 in regions 2 to 4. Eventually in the fifth region one mode quenches the other one and we have stable behavior. The experimental situations of Fig. 6 have been reproduced by a seven equation model, which couples the two complex field amplitudes (E_+ forward and E_- backward: 4 dynamical variables) via the complex second harmonic N_{2k} of the population inversion (2 more variables). The 7th real variable is the uniform component N_0 of population inversion. More explicitly, we Fourier expand the population inversion N(x) along the cell length x

$$N_k = \frac{1}{L}\int_0^L dx\; e^{-ikx} N(x) . \tag{15}$$

The 0th component is that considered in the elementary single wave approach leading to Maxwell Bloch eqs. and to the rate equations (12) (where it was called Δ). The higher components act as a diffraction grating scattering photons from the forward field into the backward and viceversa, thus providing a coupling. Our numerical solutions of the seven coupled equations for E_+, E_-, N_{2k} (complex) and N_0 (real) show that a truncation to N_{2k} is sufficient for a qualitative explanation of the experimental chaos, without having to introduce higher order components of N_k.

The physics of this chaotic ring configuration has implied a re-injection of one field into the other. We have studied specifically the single mode laser with an injected signal[12]. The injected signal has a frequency ω_1, different from the free running laser frequency ω_L. In a reference frame rotating at the frequency ω_1, the dynamical system is autonomous, but the field amplitude is made of two relevant variables, one in phase, the other in quadrature with the external signal. Adding the population difference (we refer to a CO_2 laser, hence a class B system) we have the three necessary degrees of freedom to have deterministic chaos. Indeed in ref. 12 we reported a quasi-periodic route to chaos, which in some a parameter range is described by a circle map. Further work is in progress on this line.

2. Generalized multistability and noise-induced escape from a stable attractor.

Nonlinear dissipative systems can have many simultaneously coexisting basins of attraction (generalized multistability - GM)[7,13,14]. This situation can be destabilized by changes of the control parameters , merging two independent attractors into a single one via an intermediate region which is only sporadically visited near the transition. The associated dynamics implies a low frequency tail (deterministic diffusion [15,16]). Viceversa, when the above coexistence is stable, application of external noise may induce jumps between two otherwise disjoint regions of phase space.

For some simple systems we study the two different phenomenologies near their borderline, giving a scaling relation for the escape rates in terms of control parameter and noise amplitude.

Clear evidence of generalized multistability was first shown in an electronic oscillator [13] and then in the modulated laser[7]. Fig. 7 and Fig. 8 show the coexistence of two attractors and the corresponding low frequency tails for an electronic nonlinearity obeying the Duffing equation and for the laser with modulated losses, respectively.

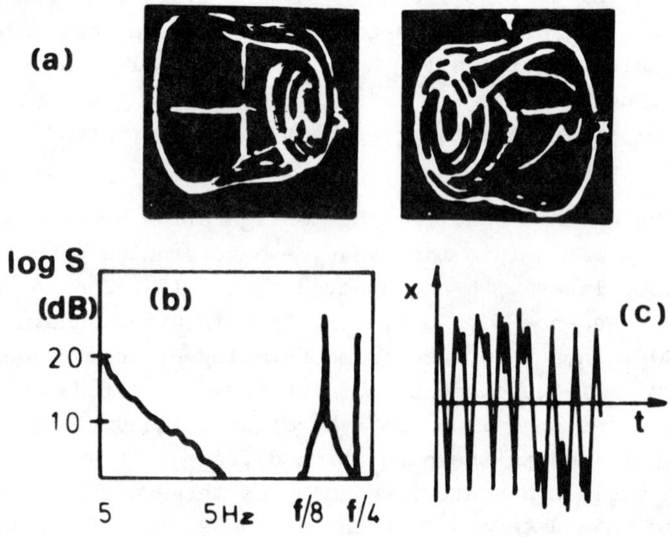

Figure 7. Hopping between two attractors and associated low frequency spectrum in the purely bistable case. (a) Symmetric phase-space plots; (b) log-log spectrum showing the low frequency branch, a broadened f/8 line, and a narrow f/4 line; (c) a sample of the x(t) plot.

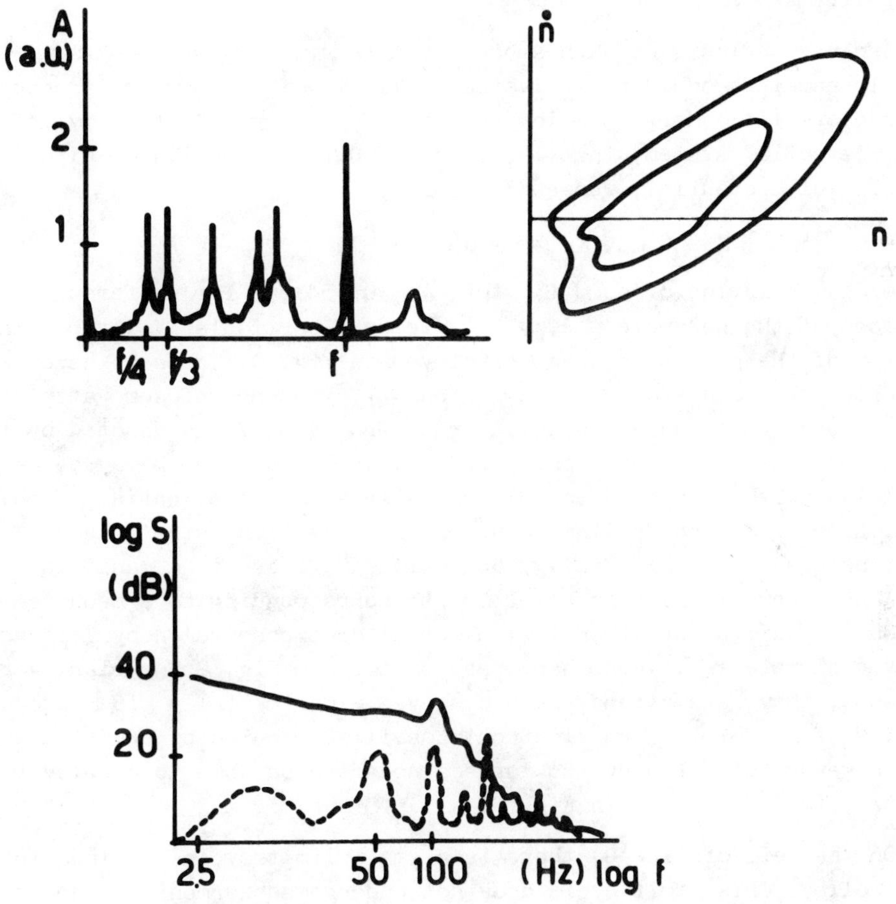

Figure 8. Bistability in a CO_2 laser with loss modulation
a-b) coexistence of two attractors (period 3 and 4 respectively)
c) comparison between the low frequency cut-off when the two attractors are stable and the low frequency divergence when the attractors are strange.

In both cases, besides the qualitative appearance of different attractors in phase space, there was a low frequency spectral component due to noise induced jumps among different attractors. Both measurements, however, might be considered as experimental artifacts. In fact, there is evidence of single attractors made of two sub-regions with infrequent passages from one to the other (see e.g. the Lorenz attractor). In such a case, the low frequency tail corresponds to the sporadic passages,and does not require added noise, (deterministic diffusion). As a matter of fact, power spectra do not permit discrimination between the two phenomena.

A clear analysis of GM has been so far limited to an oversimplified model, namely a cubic recursive map, allowing for two simultaneous attractors [14]. Here I review results of a numerical study[17] of a differential system, that is, a forced Duffing oscillator with a double well potential, whose equation is

$$\ddot{x} + \gamma \dot{x} - x + 4x^3 = A\cos\omega t. \qquad (16)$$

Numerical solution of eq. (16), for γ = 0.154 and for different choices of the parameters A, ω , yields a particularly interesting region of the parameter space displayed in Fig. 9. There we have reported the boundaries of the regions of existence of some attractors, corresponding to subharmonics of order from 4 to 7, as denoted by the associated numbers. The meaning of the two lines C and D may be explained with reference to the two valleys of the potential of eq. (16). Line C is a borderline above which the motion is no longer confined in one single valley. Below line C, there is a manifold of lines approximately parallel to it, corresponding to a sequence of subharmonic bifurcations, with mutual distances ruled by Feigenbaum universal rate δ . These have been omitted in Fig. 9 for clarity reasons. They have already been observed experimentally [18] (see Fig. 9 of Ref. 18) in an electronic oscillator ruled by eq.(16), which is, however, affected by too large a noise to be able to display the other details reported in Fig. 9.

On the left of line D, there is a small limit cycle confined in one valley. This limit cycle does not undergo subharmonic bifurcations, but it dies by intermittency on the line D. In particular, in the triangular region below the two lines C and D there is coexistence of the small limit cycle with one of the above mentioned Feigenbaum chains.

We measure the mean escape time from the period-7 attractor versus the amplitude σ of an applied noise in the region denoted by a vertical bar in Fig. 9. Specifically, at each integration step (which is 10^{-2} the forcing period) we shift $\dot{x}$ by a random number sorted from a rectangular distribution with zero mean and r.m.s. σ.

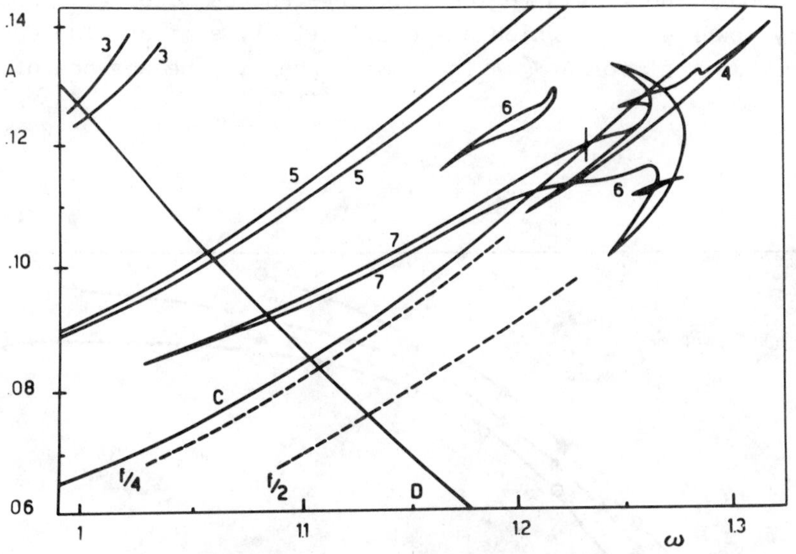

Figure 9. Phase diagram of Duffing equation showing the type of solutions for each pair of driving parameters (A, ω). The border-lines are the frontiers of parameter regions corresponding to ordered motion, and the associated numbers denote the periodicity. Curves C and D show the upper limit (in the amplitude A) for the stability region of solutions confined in one valley. The vertical bar at $\omega = 1.22$ indicates the region where the escape times have been measured.

Fig. 10 shows log T vs. $1/\sigma$ for different values (A_1, A_2, A_3, A_4) of the driving amplitude A at fixed frequency ω. $A_3 \simeq A_c$ is the parameter at which the period-7 undergoes crisis [19], that is representative points "escape" from the attractor after an infinite time, in the absence of noise. This is equivalent to the definition of crisis, i.e. A_c is the value for which the attractor of period 7 collides with the unstable period 7 solution.

For $A_4 > A_c$ even a zero noise ($1/\sigma \to \infty$) yields a finite escape time as is evident from the horizontal asymptote of the plot. For $A < A_c$ and as $\sigma \to 0$, $T \to \infty$ faster as A is reduced, as shown by comparison between A_1 and A_2. The finite escape time across a bounded region for $A > A_c$ corresponds to the deterministic (noise-free) diffusion discussed elsewhere. On the contrary, for $A < A_c$, the attractor is structurally stable, therefore, the phase point cannot escape unless we apply external noise. This is the phenomenon of noise induced jumps already reported experimentally[7, 13] and simulated with a one dimensional map[14]. Notice that, both for $A < A_c$ and $A > A_c$ around the crisis region the large escape times give low frequency power spectra which are qualitatively similar. The essential difference is that for $A < A_c$ no jumps occur in the absence of noise.

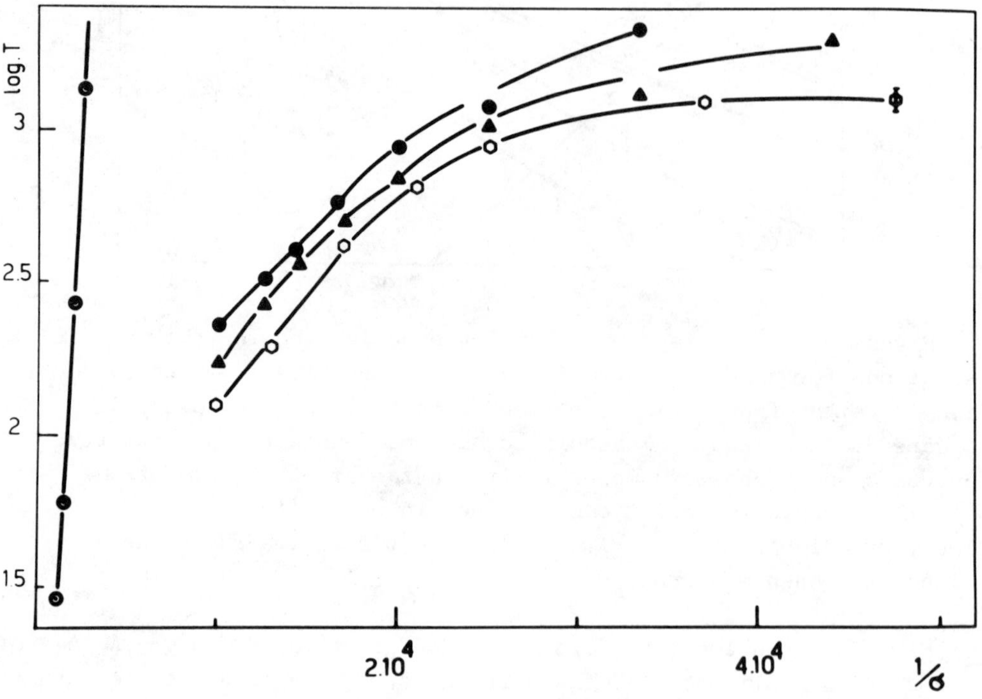

Figure 10. The mean escape time from the period-7 region vs. the inverse of the noise amplitude. All the curves refer to the same frequency, but with different A's. Namely the symbols ⊙ ● ▲ ○ represent respectively: A = .1165, .117275, .117280, .117285. The error bar is drawn only for one measurement since it is always the same.

It is apparent from Fig. 9 that one can have the same T for different A's, adjusting the noise amplitude. This may be expressed in terms of a scaling relation as given for other chaotic scenarios[20]. For simplicity we refer to the logistic map

$$x_{n+1} = \mu x_n(1 - x_n) \tag{17}$$

with μ around the crisis value $\mu_c = 4$. The invariant density for the noise-free map is given by [21]

$$\rho_c(x) = \frac{1}{\pi} \frac{1}{\sqrt{x(1-x)}} . \tag{18}$$

The divergence of ρ_c at x = 1 is due to the flat behavior of the map around the pre-image point x = 1/2.

We first evaluate the mean escape rate R from the right for $\mu \sim \mu_c$. If, either a small noise is added, or μ is set slightly above μ_c, there is no longer an invariant density; however, we can reasonably assume that the distribution of points relaxes to a pseudo-invariant $\rho(x)$ on a time scale much shorter than the mean escape time. In particular, for $\mu < \mu_c$ and without noise, $\rho(x)$ may be a highly singular distribution, however an additional random term smooths out almost all of the peaks over a continuous $\rho(x)$, whose rightmost part is a scaled version of the corresponding part of $\rho_c(x)$, that is, $\rho(x) \simeq 1/\pi\sqrt{1-\epsilon-x}$, with the asymptote shifted from x = 1 to $x = 1 - \epsilon$ ($\epsilon = 1 - \mu/4$).

When applying a Gaussian noise of r.m.s. σ. the escape rate R at the point x is equal the area of the Gaussian centered at x, outside the segment (0,1). Hence, for $\epsilon \ll 1$, we have the escape rate from right given by ($y = 1 - x - \epsilon$)

$$R(\epsilon, \sigma) = \frac{1}{\pi} \int_0^\infty \frac{dy}{\sqrt{y}} \int_y^\infty \frac{e^{-\frac{(z+\epsilon)^2}{2\sigma^2}}}{\sqrt{2\pi\sigma^2}} dz . \tag{19}$$

Integrating by parts we have [22]

$$R(\epsilon, \sigma) = \frac{1}{\pi} \sqrt{\frac{\sigma}{2}}\, e^{-\epsilon^2/4\sigma^2} D_{-3/2}(\epsilon/\sigma), \tag{20}$$

where D is the parabolic cylinder function.

It is readily seen that the function R (ϵ, σ) satisfies the general scaling law

$$R(\epsilon, \sigma) = \sigma^\alpha F(\epsilon/\sigma^\beta), \tag{21}$$

with the exponents $\alpha = 1/2$ and $\beta = 1$. Similar scaling laws have already been given for the period doubling cascade and for intermittency [20], but with different exponents, as well as a different physical meaning for R, as discussed later in the conclusion.

At the crisis ($\epsilon = 0$) R scales as the square root of σ, namely,

$$R(0,\sigma) = \frac{\Gamma(3/4)}{\pi^{3/2}} 2^{1/4} \sigma^{1/2} . \tag{22}$$

Below the crisis ($\epsilon > 0$) and for $\epsilon \gg \sigma$, we have the asymptotic relation

$$R(\epsilon,\sigma) = \frac{\sqrt{2}}{\pi} \frac{\sigma^2}{\epsilon^{3/2}} e^{-\epsilon^2/2\sigma^2} . \tag{23}$$

Above the crisis ($\epsilon < 0$) we can escape even without noise: indeed the asymtotic expansion of eq.(20) in the limit of large ϵ/σ yields $R \approx (2/\pi)|\epsilon|^{1/2}$ as already given in Ref. 19.

To evaluate the left escape rate L, we start again from the crisis density which has the same divergence at x = 0 as at x = 1. However, while the divergence at x = 1 is provided by injection from an inner region, the divergence at x = 0 is a straightforward consequence of the density around its pre-image x = 1 and hence strongly affected even by a small noise.

We show that L is much smaller (around 10%) than R at $\mu = \mu_c$ and even less important for $\mu < \mu_c$ for noise induced jumps.

For $\mu > \mu_c$ the noiseless escape always occurs at right (image of points around the maximum). Numerical studies show that, both for Gaussian and rectangular noise, the left contribution is around 10% (namely 13% for the former and 11.5% for the latter) of the right one. Hence, we consider it sufficient to develop a handy argument for L based on the application of a rectangular noise r.m.s. σ. At the crisis, the density $\rho(x) = (\pi\sqrt{1-x})^{-1}$ around x = 1, once convoluted with the rectangular noise, yields a distribution

$$P(x) = \frac{1}{\pi\sigma\sqrt{3}} \sqrt{1+\sqrt{3}\sigma - x} , \quad 1-\sqrt{3}\sigma < x < 1+\sqrt{3}\sigma, \tag{24}$$

whose area on the right of 1 directly gives the escape rate $\bar{R}$

$$\bar{R}(0,\sigma) = \frac{2}{3^{3/4}} \sigma^{1/2} . \tag{25}$$

Incidentally, we can observe that the rectangular noise yields the same scaling relation as the Gaussian one (see eq.(21)) with the only difference of multiplicative factor 7% larger in eq. (24).

By applying the recursive map (17), we get a first approximation of the pseudo invariant ρ (x) around x = 0

$$\rho(x) \approx (1/8\sqrt{3}\pi\sigma)\sqrt{x + 4\sqrt{3}\sigma} , \qquad 0 < x < 4\sqrt{3}\sigma . \tag{26}$$

Now, a convolution of eq. (26)with the noise gives a distribution P(x) whose area on the left of 0 yields a first order contribution to L. Furthermore, by applying again the map, since x = 0 is an unstable fixed point, the remaining area of P is partially shifted away . Hence, we get a second order contribution to L by taking the convolution with the noise. The procedure is rapidly convergent. Indeed, each contribution is about 7 times smaller than the previous one. The sum of the first three terms is

$$\bar{L} = 3.155\ 10^{-2}\ \sigma^{-1/2} \qquad (27)$$

yielding a theoretical prediction of the ratio $\bar{R}/\bar{L} \sim 8.85$. This is in agreement with the numerically estimated value 8.69 ± 0.09. Now, by summing up the left and right contributions we are able to evaluate the global mean escape time for the interval (0,1). A good agreement between the numerical and theoretical results is shown in Fig.11.

In conclusion, we have analysed the effect of an external noise in a dynamical system near a crisis. We have described the noise-induced jumps by means of mean escape times from the attractor. An analytic expression for a scaling law has been found showing a strict analogy with the period-doubling and the intermittency phenomenon. While in the doubling the relevant quantity is the Lyapunov exponent and in the intermittency it is the laminar length, here the major role is played by the mean first-passage time.

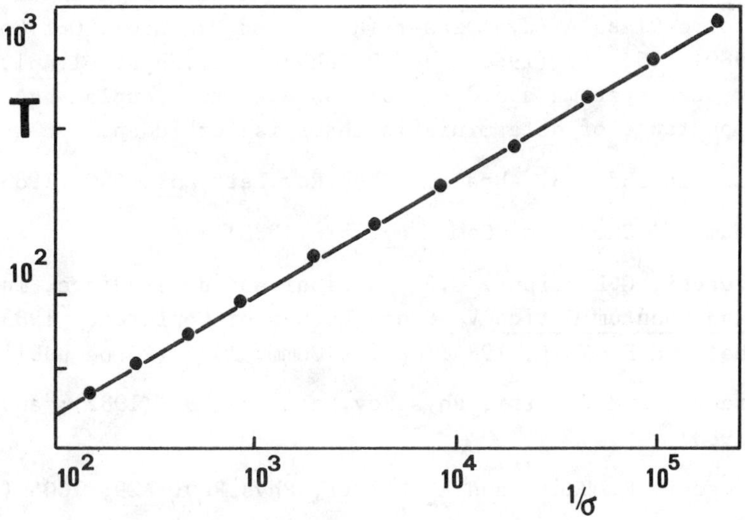

Figure 11. The mean escape time from the interval (0,1) vs. the inverse of the noise amplitude for numerical integration of the logistic map at μ = 4 and with a Gaussian noise. The straight line shows the theoretical prediction.

REFERENCES

1 F.T. Arecchi, "Photocount distributions and field statistics" (Proc.E.Fermi School 1967) in Quantum Optics ed. by R.J. Glauber, Academic Press, 1969

2 H. Haken, Synergetics, Springer Verlag, 1977

3 F.T. Arecchi, "Experimental Aspects of Transition Phenomena in Quantum Optics" (Proc. XVIIth International Solvay Conference on Physics, 1978) in Order and Fluctuations in Equilibrium and Non-equilibrium statistical Mechanics ed.by G. Nicolis et al., J. Wiley, 1981

4 S. Chandrasekhar, Hydrodynamics and Hydromagnetic Stability, Oxford U., London 1961

5 E. Lorenz, J. Atmos. Scie. 20, 130 (1963)

6 H. Haken, Phys. Lett. 53A, 77 (1975)

7 F.T. Arecchi, R. Meucci, G. Puccioni, J. Tredicce, Phys.Rev.Lett. 49, 1217 (1982)

8 J. Tredicce, G.L. Lippi, F.T. Arecchi and N.B. Abraham, Proc.Royal Soc. (London), (to be published);
N.B. Abraham, F.T. Arecchi, G.L. Lippi and J. Tredicce, Infrared Physics, (to be published)

9 Oscillatory instabilities in a ring cavity have been shown previously for a Class A (dye) laser (R.Roy and L.Mandel, Opt.Comm. 34, 133 (1980); 35, 247 (1980), and D.Kühlke, Act.Phys.Pol.A61, 547 (1982)). Such systems are described by only two coupled equations, thus appearance of deterministic chaos is forbidden.

10 R.S. Gioggia and N.B. Abraham, Phys.Rev.Lett. 51, 650 (1983)

11 R. Graham, Y. Cho, Opt. Comm. 47, 52 (1983)

12 F.T. Arecchi, G.L. Lippi, G.P. Puccioni and J. Tredicce, in Coherence and Quantum Optics V (Proc. Rochester Conference 1983) ed. L. Mandel and E. Wolf, 1984; Optics Comm. 1984 (to be published)

13 F.T. Arecchi and F. Lisi, Phys.Rev.Lett. 49, 94 (1982); and 50, 1328 (1983)

14 F.T. Arecchi, R. Badii and A. Politi, Phys.Rev. A29, 1006 (1984)

15 T. Geisel and J. Nierwetberg, Phys.Rev.Lett. 48, 7 (1982);
S. Grossman and H. Fujisaka, Phys.Rev. A26 , 504 (1982)

16 Y. Aizawa, Prog. Theor. Phys. 68 , 64 (1982)

17 F.T. Arecchi, R. Badii, and A. Politi, Phys. Lett. 103A, 3 (1984)

18 F.T. Arecchi and A. Califano, Phys. Lett. 101A, 443 (1984)

19 C. Grebogi, E. Ott and J.A. Yorke, Phys. Rev. Lett. 49, 1507 (1982)

20 J. Crutchfield, M. Nauenberg, J. Rudnick, Phys.Rev.Lett. 46, 933 (1981); R. Shraiman, C.E. Wayne and P.C. Martin, Phys. Rev. Lett. 46, 935 (1981); J.G. Hirsch, M. Nauenberg, D.J. Scalapino, Phys. Lett. 87A, 391 (1982)

21 P. Collet and J.P. Eckmann, Interated Maps on the Interval as Dynamical Systems (Birkhäuser, Boston 1980)

22 I.S. Gradshteyn and I.M. Ryzhik, Table of Integrals Series and Products (Academic Press, New York 1965)

CHAOS AND MULTIPLE PHOTON ABSORPTION

JAY ACKERHALT* AND P. W. MILONNI**

Abstract. An anharmonic vibrational mode of a molecule, driven by an intense infrared laser and coupled to a quasi-continuum of background modes, is found to undergo chaotic oscillations. This chaos leads to predominantly fluence-dependent rather than intensity-dependent multiple-photon absorption, as is found experimentally. The loss of "coherence" is associated with the decay of temporal correlation of background-mode oscillations.

1. Introduction. For some time now is has been known that molecules like SiF_4 or SF_6 dissociate rather easily in infrared laser fields. Initially this was something of a surprise, for the vibrational anharmonicity ruins the equal level spacing of the ideal harmonic oscillator, making the molecule off-resonant after the absorption of a few photons. Furthermore the multiple-photon absorption process is found to be predominantly fluence-dependent rather than intensity-dependent; that is, the process depends strongly on the total energy of the laser pulse, but not so much on the pulse intensity.[1,2] It is our opinion that neither of these two features of multiple-photon excitation (mpe) have been very satisfactorily explained, in spite of many attempts during the past 10-15 years. In this paper a novel explanation is offered.[3]

The basic model of mpe considered here is outlined in Section 2. Details may be found elsewhere.[4] In Section 3 we discuss evidence of chaotic behavior. We find that the energy deposited in the molecule is proportional to time, as has been effectively assumed in earlier attempts to model the mpe experiments. In Section 4 we discuss further the model and its implications, and close with some questions for further investigation.

2. The model. Here we consider only the essential parts of the model in order to see how chaotic behavior arises and how it leads

*Los Alamos National Laboratory, Los Alamos, NM 87545.
**Department of Physics, University of Arkansas, Fayetteville, AR 72701. Research supported by NSF grant PHY-8308048, and by dual funding from NSF EPSCOR grant ISP-8011444 and the State of Arkansas.

to strongly fluence-dependent mpe. The most important restriction will be the neglect of molecular rotations, which have been considered elsewhere in connection with chaotic dynamics.[4,5] The model Hamiltonian is

$$H = H_{PM} + H_{BM} + H_{IC} + H_{PMF} \tag{1}$$

where H_{PM} is the Hamiltonian for an infrared-active pump mode, taken to be

$$H_{PM} = \Delta a^{\dagger} a - \chi (a^{+} a)^{2} \tag{2}$$

where a is the lowering operator for vibrational quanta, Δ is the detuning of the laser field from (harmonic) resonance, and χ is the molecular anharmonicity; [6] H_{BM} is similarly the Hamiltonian for N background modes, which are assumed not to be coupled to the laser and furthermore to be perfectly harmonic:

$$H_{BM} = \sum_{m=1}^{N} (\Delta + \varepsilon_m) b_m^{\dagger} b_m \tag{3}$$

where $\Delta + \varepsilon_m$ is the energy associated with the vibrational quanta of mode m, and b_m is the corresponding lowering operator; H_{IC} describes the coupling of pump and background modes, assumed to occur via one-quantum exchanges:

$$H_{IC} = \sum_{m=1}^{N} \beta_m (a^{\dagger} b_m + b_m^{\dagger} a) \tag{4}$$

and H_{PMF} describes the coupling of the pump mode to the laser, also assumed to occur via one-quantum exchanges:

$$H_{PMF} = \frac{\Omega}{\sqrt{n}} (c^{\dagger} a + a^{\dagger} c) \tag{5}$$

Here Ω is basically the "Rabi frequency" determined by the product of the applied electric field and the transition dipole moment.[7] c is the photon annihilation operator, and for later convenience we have separated out the "photon number" n of the applied field. This is only a formal convenience, as the field will be treated purely classically in the calculations. From the Hamiltonian (1) it follows that the total excitation number

$$\eta = a^{\dagger} a + c^{\dagger} c + \sum_{m} b_m^{\dagger} b_m \tag{6}$$

is a conserved quantity.

The problem is considerably simplified under certain assumptions and approximations. We assume first $\beta_m = \beta$ for all m. As in radiationless transition theory it is also convenient to assume that the background mode frequencies are uniformly distributed [8]:

(7) $$\varepsilon_m = \Delta_0 + m\rho^{-1}$$

where Δ_0 is the separation from the pump mode of the nearest background mode, and ρ is the density of background states. Finally we assume an infinite number of background modes above and below the pump mode frequency. This last assumption is considerably more reasonable than one might at first suppose, because only a few tens of background modes are typically necessary to simulate the $n\to\infty$ limit.[8,9] This treatment of the background is often called the "quasi-continuum model."

In addition to the quasi-continuum approximation we make one fundamental assumtion, namely that it is reasonable to treat the entire dynamics classically. This is surely a good approximation (for our purposes) for the laser field, because we are dealing with a very large number of incident photons. It is not as easy to justify the classical approximation to the molecular vibrations. However, mpe typically involves the absorption of perhaps 30 photons, making multiphoton resonance unimportant, and so we can expect the classical vibrator model to be an acceptable approximation. Other researchers, particularly Lamb, have advocated such a classical approach to mpe studies.[10,11]

In the classical approximation a, b_m, and c are no longer operators but just ordinary functions of time. The number of photons absorbed from the field by the molecule is easily shown to be [3,4]

(8) $$n - |c(t)|^2 = -2\Omega \int_0^t dt' \, \mathrm{Im}[a(t')]$$

The quasi-continuum approximations lead to the following delay differential equation for the complex pump-mode amplitude a(t) [3,4]:

(9) $$\dot{a}(t) = -i(\Delta - \frac{i\gamma}{2})\, a(t) + 2i\,\chi\, a(t)|a(t)|^2 - i\Omega - \gamma \sum_{m=1}^{\infty} \bar{\beta}^m\, a(t-m\tau)\,\theta(t-m\tau)$$

where θ is the Heaviside step function and

(10a) $$\gamma = 2\pi\beta^2\rho$$

(10b) $\tau = 2\pi\rho$

(10c) $\bar{\beta} = e^{-i(\Delta+\Delta_0)\tau}$

Note that γ is just the Fermi Golden Rule rate that would be predicted by (second-order) perturbation theory. τ is a "recurrence time" that appears as a consequence of the uniform spacing of the background modes; formally it arises from the Poisson summation formula.[3,4]

Equations (8) and (9) together describe the mpe process. From equation (9) it is possible to derive a discrete mapping under certain approximations.[4]

3. Chaotic Dynamics. Using realistic models of the laser and molecular parameters Δ, Δ_0, γ, χ, and Ω, we have found that the dynamical system (9) is chaotic. This has been inferred both from power spectra and the computation of (positive) Lyapunov exponents.[3,4]

The appearance of this chaotic behavior may well be responsible for the "incoherent," fluence-dependent mpe observed experimentally and noted earlier. To better appreciate this we consider the total "excitation number" in the background modes, which may be written

(11) $$\sum_m |b_m(t)|^2 = \gamma \int_0^t dt' |a(t')|^2 + 2\mathrm{Re} \sum_{m=1}^{\infty} \bar{\beta}^m \theta(t-m\tau) \int_0^{t-m\tau} dt'\, a^*(t')a(t'+m\tau) \quad .$$

The second term involves the time-averaged correlation function $\langle a^*(t)a(t+T)\rangle$, which dies out with increasing T, just as the corresponding correlation functions vanish for the logistic mapping and the Lorenz model. The major contribution to (11) is therefore from the first term on the right. This term describes, in a simple rate-equation fashion, the growth of background-mode excitation. The growth rate is simply $\gamma|a|^2$, where $|a|^2$ is the pump mode excitation and γ is the Fermi Golden Rule rate. It should be emphasized, however, that we have not used perturbation theory, nor have we needed to assume a dense distribution of final states. The "rate-equation approximation" emerges instead as a consequence of chaos and the associated decay of temporal correlation, and is applicable even if the background mode density is relatively sparse.

The absorption of energy from the laser is found by integration of (9), using (8). A typical sort of result is shown in Figure 1. In gross terms the absorption is linear with time, implying approximately fluence-dependent mpe. The background modes act as a sponge for the

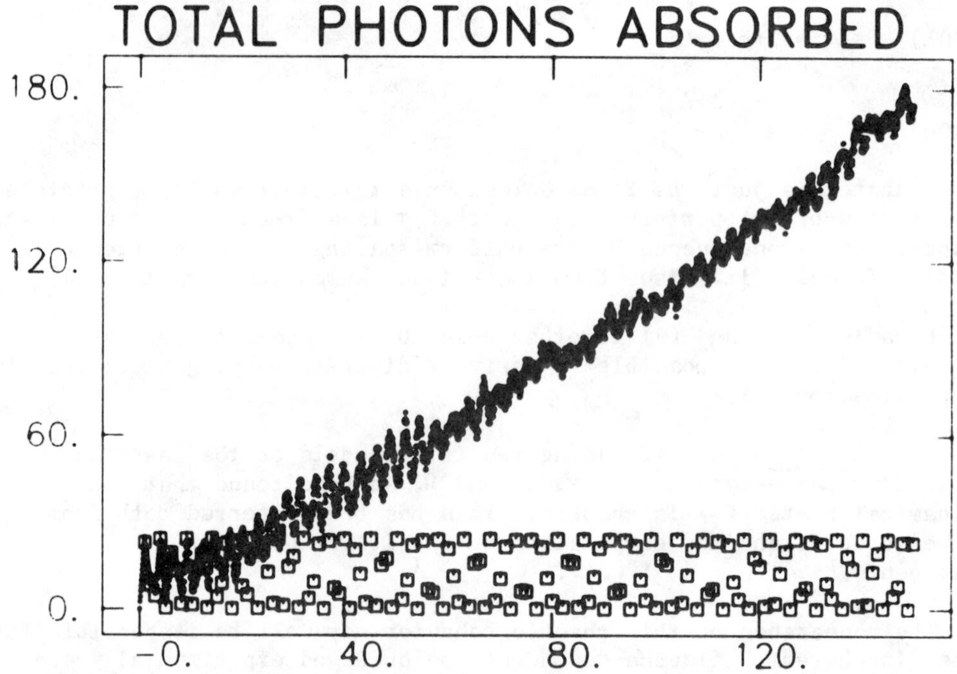

Figure 1. A typical computed curve of photons absorbed vs. time, showing approximately linear dependence on time. For appropriate parameter ranges in these computations see Reference [4].

laser energy, soaking up energy at a rate $\gamma|a|^2$, according to the discussion above; this interpretation is corroborated by numerical experiments showing that the first term is the dominant contribution to (11).

4. Discussion. We have advocated here the view that fluence-dependent absorption can be predicted without a statistical, rate-equation treatment of the background "quasi-continuum." In particular, a large density of background states is not required. As present estimates of background densities suggest that they are not large enough to justify a statistical treatment, chaos emerges as a very plausible explanation of fluence-dependent, collisionless mpe. The "incoherence" of the observed process is found without having to assume coherence-destroying damping terms.

The dependence of multiple-photon absorption on pulse energy rather than intensity has been found for about 50 polyatomic molecules by Lyman, et al.[1,2] It has usually been assumed that "hotband" and rotational averaging are rsponsible for this trend. However, molecular rotation effects cannot necessarily be treated by a simple averaging process.[5]

In summary, chaotic dynamics of molecular vibrations can explain the strongly fluence-dependent nature of collisionless, infrared laser mpe. The fact that the background modes act as a sponge for the laser energy may also help to explain the difficulty encountered so far in realizing mode-selective laser chemistry. Chaos is found in a fairly realistic model of mpe.

A somewhat similar "stochastic mechanism" of mpe has been considered by Belobrov, et al.[12] Their "stochastic mechanism" (chaos), however, requires the molecules to interact with their collective radiation field.

One important question we have not addressed in any quantitative fashion here is the implication of quantum effects. A system like the one considered would be expected to have discrete (quantum-mechanical) energy levels, and so the question arises of how the classical chaos, with its broadband power spectrum, manifests itself quantum mechanically. In particular, can we expect our explanation of fluence-dependent mpe to survive in a quantum-mechanical treatment of the problem?

REFERENCES

[1] J. L. Lyman, G. P. Quigley, and O. P. Judd, Single Infrared Frequency Studies of Multiple-Photon Excitation and Dissociation of Polyatomic Molecules, in Multiple-Photon Excitation and Dissociation of Polyatomic Molecules, edited by C. D. Cantrell, Springer-Verlag, N.Y., 1984.

[2] O. P. Judd, Quantitative Comparison of Multiple-Photon Absorption in Polyatomic Molecules, J. Chem. Phys. 71 (1979), pp. 4515-4530.

[3] J. R. Ackerhalt, H. W. Galbraith, and P. W. Milonni, Chaos in Multiple-Photon Absorption of Molecules, Phys. Rev. Lett. 51 (1983), pp. 1259-61.

[4] J. R. Ackerhalt, P. W. Milonni, and M.-L. Shih, Chaos in Quantum Optics, Physics Reports, to be published.

[5] H. W. Galbraith, J. R. Ackerhalt, and P. W. Milonni, Chaos in the Multiple Photon Excitation of Molecules Due to Vibration-Rotation Coupling at Lowest Order, J. Chem. Phys. 79 (1983), pp. 5340-50.

[6] G. Herzberg, Infrared and Raman Spectra of Polyatomic Molecules, Van Nostrand Reinhold, N.Y., 1945.

[7] L. Allen and J. H. Eberly, Optical Resonance and Two-Level Atoms, Wiley Interscience, 1975.

[8] M. Bixon and J. Jortner, Intramolecular Radiationless Transitions, J. Chem. Phys. 48 (1968), pp. 715-726.

[9] P. W. Milonni, J. R. Ackerhalt, H. W. Galbraith, and M.-L. Shih, Exponential Decay, Recurrences, and Quantum-Mechanical Spreading in a Quasi-Continuum Model, Phys. Rev. A. 28 (1983), pp. 32-39.

[10] W. E. Lamb, Jr., Multiphoton Dissociation of Polyatomic Molecules: Quantum or Classical?, in Laser Spectroscopy III, edited by J. L. Hall and J. L. Carlsten, Springer-Verlag, N.Y., 1977.

[11] W. E. Lamb, Jr., Classical Model of SF_6 Multiphoton Dissociation, in Laser Spectroscopy IV, edited by H. Walther and K. W. Rothe, Springer-Verlag, N.Y., 1979, pp. 296-299.

MOTION OF QUANTUM SYSTEMS UNDER EXTERNAL TIME-PERIODIC PERTURBATION

GIULIO CASATI* AND ITALO GUARNERI**

Abstract. The chances that chaos has to survive in quantum mechanics are briefly discussed. The significant limitations that quantum mechanics imposes on classical chaotic motion are examined for the problem of excitation and ionization of hydrogen atoms by microwave field. For this problem, a direct experimental check on the theoretical results should be possible.

The richness and the variety of the classical motion, already foreseen by Poincaré at the start of this century, has by now fully revealed its fundamental relevance, for the understanding of a wide class of phenomena, in physics as well in other disciplines, that can be described in classical i.e., non quantum terms.

One of the main discoveries of the last years was that deterministic classical systems, may exhibit a type of motion which is completely indistinguishable from a truly random one. The precise meaning is that almost all trajectories are random in the sense of algorithmic complexity theory [1]; namely, they are unpredictable or uncomputable. For these orbits a sequence of measurements with finite precision performed up to a certain time, will never provide sufficient information to predict the outcomes of later measurements. In other words, in order to predict the solution one has to give it in some way beforehand. This, in essence, is the reason why it is impossible to make long-term weather predictions; in fact, the length of the shortest algorithm that can predict the state of weather after a time t is asymptotically proportional to t; hence, this shortest algorithm is in a sense no more convenient than waiting a time t and observing the weather. On the other hand the above proper-

*Dipartimento di Fisica, Università di Milano, Via Celoria 16,20133 Milano, Italy

**Dipartimento di Fisica Nucleare e Teorica, Università di Pavia, Via Bassi, 27100 Pavia, Italy

ties of motion provide the possibility of deriving classical statistical mechanics without additional assumptions and/or recourse to the so-called thermodynamic limit. For example, the Fourier law of heat conduction has been recently verified [2] in a purely dynamical deterministic system.

Turning now to quantum mechanics, we are faced at once with a striking difference. Despite the fact that quantum mechanics is an essentially statistical theory, there is nothing, in the type of evolution described by the Schrödinger equation, so complex as the structure of classical orbits. Indeed, the discreteness of the energy spectrum of finite particle, conservative quantum systems leads to almost periodicity in time of the motion so that at most, ergodic behaviour is allowed, which is very far from the kind of unpredictability sketched above. Still, it is possible that the complexity of the classical motion reappears here in some particular feature of the energy level statistics, or of the distribution of eigenfunctions. One may inquire for example whether it is possible or not to distinguish between integrable and chaotic systems just by looking at the sequence of eigenvalues or at the spatial structure of eigenfunctions. This subject is now under intense investigations, aimed at verifying whether a difference in complexity of classical systems has a definite counterpart in some difference of complexity of their spectral sequences. We shall only mention here a recent result[3] that for a simple class of integrable systems the sequence of eigenvalues has zero algorithmic complexity; in particular, it is not a random sequence. As to classical chaotic systems, no definite results concerning the degree of randomness of their spectral sequences is so far known.

Perhaps more illuminating is the consideration of systems under external periodic perturbations since they may allow for a continuous spectrum of the motion (the so-called quasi-energy spectrum). A particular example in this class of systems which has been extensively studied [4-9] , is the so-called δ-kicked rotator - a rotator subject to external time-periodic pulses. Although a number of very interesting properties have been found, it is not yet completely clear whether this quantum system can exhibit some kind of disordered motion. Certainly it is not so squarely almost-periodic as conservative systems: in fact, it has been recently shown [8] that , as the external perturbation is turned on a transition from a pure point to a partly continuous spectrum occurs, which presents interesting similarities to the phenomenon described by the KAM theory of classical mechanics. Just as for Arnold's diffusion, it is not yet clear whether the motion associated with this continuous component has any "chaotic" feature; anyway, on account of numerical experiments, we can say that, in this case also, quantization places severe limitations to the classical chaotic motion; this feature seems to be imputable to the singular, rather than absolute continuity, of the quasi-energy spectrum which does not allow for a fully chaotic motion; we can expect, at best, some kind of weak mixing.

If the above limitation is of a general nature, then it should be possible to devise experiments which give evidence of this lack of chaotic behaviour in quantum systems. In order to illustrate the actual possibility of such experiments, we shall now briefly describe a more realistic system, which provides a model for the behaviour of an hydrogen atom in a microwave field. In recent experiments [10-11] the ionization rate of highly-excited hydrogen atoms by a microwave monochromatic field has been measured. Even though the microwave frequency ($\sim$ 10GHz) was about one half of the frequency necessary for the resonant transition from the initial level $n_0=66$ to the next level n=67, nevertheless a noticeable ionization was detected for a microwawe field intensity much below the classical ionization threshold and also of the quantal tunneling threshold. Due to the high quantum numbers, a classical description may seem appropriate and indeed in ref. [12] the relevance of chaotic motion in the diffusion mechanism leading to ionization has been stressed.

A full quantum numerical study has been given in refs. [13-14] in which the hydrogen atom excitation was studied starting with initial states very extended along the field direction, i.e. states with parabolic quantum numbers $n_1 >> n_2 \sim 1$ and m=o. This situation is conveniently described by the one-dimensional Hamiltonian

$$H = \frac{P_z^2}{2} - \frac{1}{|z|} + \varepsilon z \cos \omega t \tag{1}$$

where ε, ω are the field strenght and frequency in atomic units. Hamiltonian (1) was used in refs. [15] to study surface state electrons bounded to the surface of liquid Helium by their image charge (in this case there is also an infinite repulsive barrier at the surface, which is due to the Pauli exclusion principle).

The quantum limitation of chaotic diffusion was observed also here as it is clearly appearent from Fig. 1 in which the second moment

$M_2 = \frac{\langle (n - \langle n \rangle)^2 \rangle}{n_0^2} \equiv \frac{\langle (\Delta n)^2 \rangle}{n_0^2}$ is plotted versus time $\tau = \omega t/2\pi$

for typical cases $n_0 = 66$, $\omega_0 = \omega n_0^3 = 1.2$, $\varepsilon_0 = \varepsilon n_0^4 = 0.03$ and 0.04. In both cases the increase in the quantum moment M_2 is much smaller than in the classical situation .

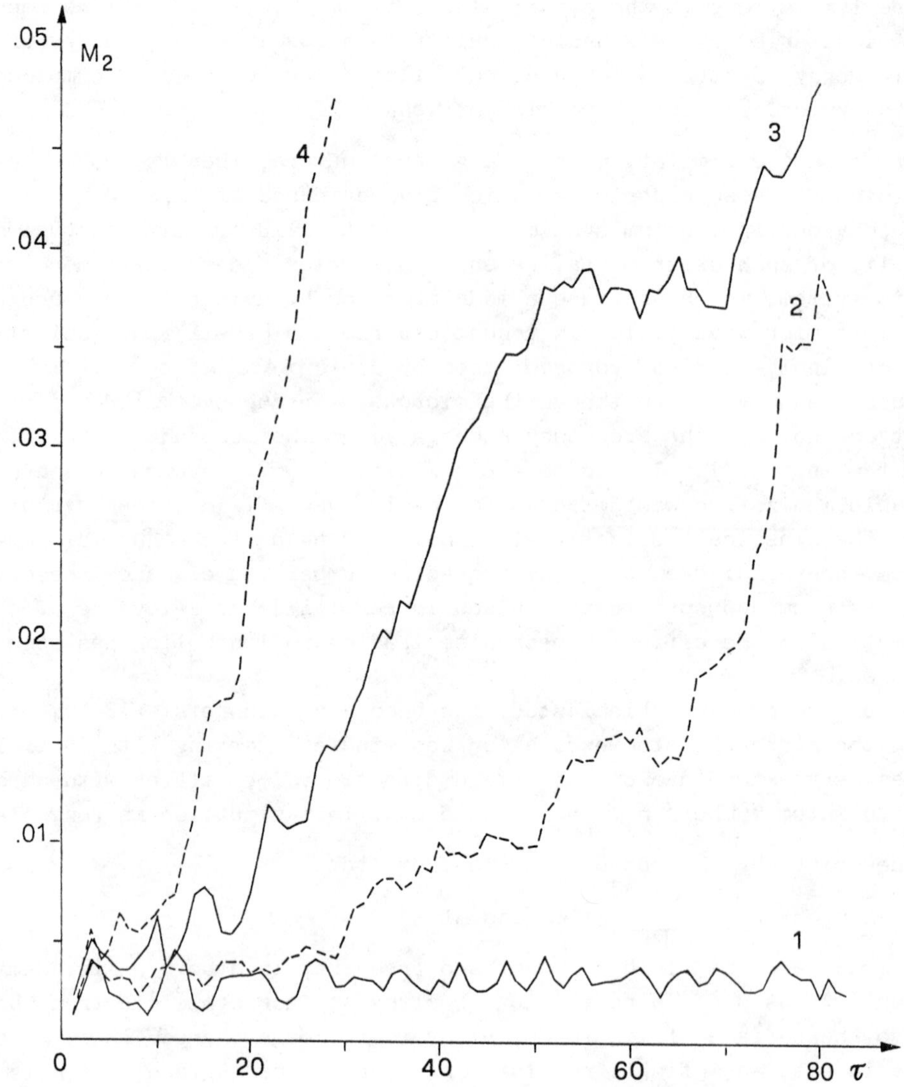

Figure 1. Dependence of the second moment $M_2 = \langle (n-\langle n\rangle)^2\rangle / n_0^2$ on time $\tau = \omega t/2\pi$ measured in the number of field periods. Quantum case: $n_0=66$, $\omega_0 = 1.2$; $\varepsilon_0= 0.03$ (curve 1), $\varepsilon_0 = 0.04$ (curve 3). Curves 2 and 4 correspond to the classical limit of 1 and 3 respectively.

In Fig. 2 we plot the distribution function $\bar{f}_n(\tau)$ (probability of occupation of the n-th level at time τ) averaged over 40 values in the interval $80 < \tau \leq 120$ for the case $n_0 = 66$, $\omega_0 = 1.2$, $\varepsilon_0 = 0.03$. Notice that the packet remains localized around the initial value $n = n_0$, except for a multiphoton "plateau" with a definite resonant

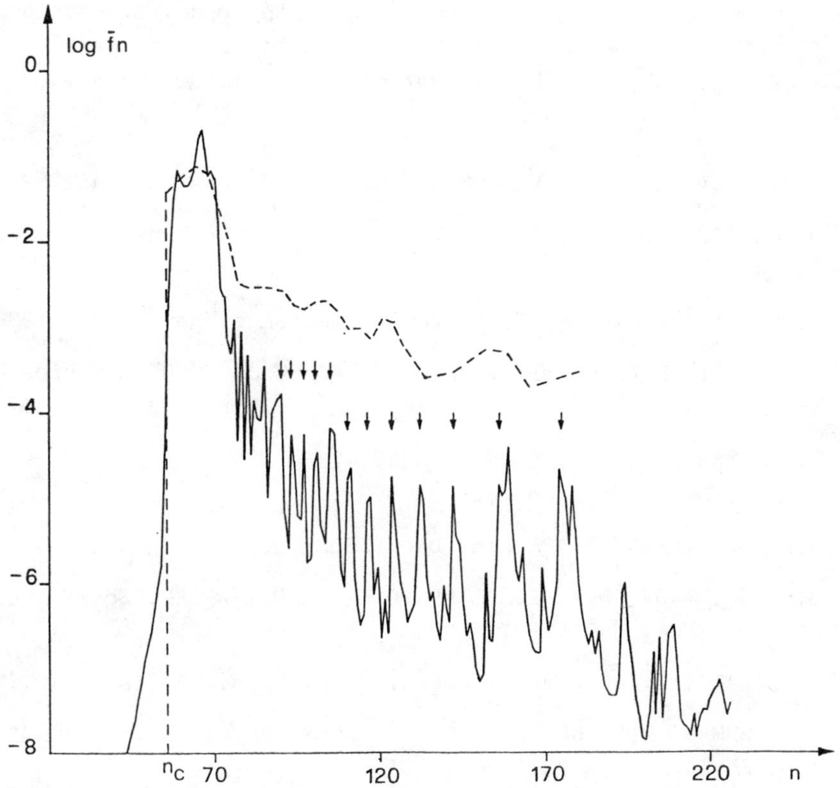

Figure 2. Distribution function $\bar{f}_n$ averaged over 40 values within the interval $80 < \tau < 120$ for the quantum case 1 of Fig. 1 (full line) and for the classical case 2 of Fig. 1 (dashed line). n_c is the classical chaos border. Arrows are drawn with equal spacing $\Delta E = \omega$ (in energy scale), one arrow being attached to the empirical peak at n = 142.

structure with peaks equally-spaced (in energy) at the frequency of the external field.

On the other hand, when the intensity of the microwave field exceeds a critical value $\epsilon_q \simeq \omega_0^{7/6}/2n_0^{1/2}$, called the <u>quantum delocalization border</u> [14] the quantum momentum M_2 keeps growing and indefinite diffusion seems to take place.

This type of motion naturally presents itself as a candidate for the quantum equivalent of classical diffusion even though, from a theoretical viewpoint we should clarify what degree of disorder or irreversibility is actually present in it, before identifying it with "quantum diffusion".

Even more interesting would be to check the results described above in a direct laboratory experiment. This would, for the first time, reveal a manifestation of chaotic motion in a quantum system.

REFERENCES

[1] V.M. ALEKSEEV, M.V. YAKOBSON, Physics Reports 75 (1981) 287.

[2] G. CASATI, J. FORD, F. VIVALDI, W.M.VISSCHER, Phys. Rev. Lett. 52 (1984) 1861.

[3] G. CASATI, I. GUARNERI, F. VALZ-GRIS, Phys. Rev. A 30 (1984) 1586.

[4] G. CASATI, B.V. CHIRIKOV, F.M. IZRAILEV, J. FORD, Lectures Notes in Physics Vol. 93 ed. G. Casati and J. Ford, Springer 1979.

[5] F.M. IZRAILEV, D.L. SHEPELYANSKY, Theor. Mat. Fiz. 43 (1980) 417 (Theor. Math. Phys. 43 (1980) 553).

[6] D.L. SHEPELYANSKY, Physica D 8 (1983) 208.

[7] B.V. CHIRIKOV, F.M. IZRAILEV, D.L. SHEPELYANSKY, Soviet Scientific Reviews, sec. C 2 (1981) 209.

[8] G. CASATI, I. GUARNERI, Commun. Math. Phys. 95 (1984) 121.

[9] S. FISHMAN, D.R. GREMPEL, R.E. PRANGE. Phys. Rev. Lett. 49 (1982) 509; Phys. Rev. A 29 (1984) 1639.

[10] J.E. BAYFIELD, P.M KOCH Phys. Rev. Lett. 33 (1974) 258.

[11] J.E. BAYFIELD, L.D. GARDNER, P.M. KOCH Phys. Rev. Lett. 39 (1977) 76.

[12] B.I. MEERSON, E.A. OKS, P.V. SASOROV. Pis'ma Zh. Eksp. Teor. Fiz., 29 (1979) 79.

[13] D.L. SHEPELYANSKY, preprint 83-61, INP, Novosibirsk, 1983; Proc. Int. Conf. on Quantum Chaos, Como 1983 (Plenum, 1984).

[14] G. CASATI, B.V. CHIRIKOV, D.L. SHEPELYANSKY, Quantum limitations for chotic excitation of hydrogen atom in monochromatic field". Preprint.

[15] R.V. JENSEN , Phys . Rev. Lett. 49 (1982) 1365; Phys. Rev. A 30 (1984) 386.

FINITE DIMENSIONAL ASPECTS OF TURBULENT FLOWS

O. P. MANLEY*

Abstract. It is demonstrated how the conventional estimate of the number of degrees of freedom of turbulent flows, based on purely physical and dimensional arguments, can be obtained from the properties of the Navier-Stokes equation. Also the dependence of the sufficient number of degrees of freedom and of the dimension of the attractor on the Reynolds number is elucidated.

1. Introduction. Recent abstract research on the asymptotic properties of Navier-Stokes equations should prove valuable to the furtherance of the use of computers as experimental tools in the study of the dynamics of fluids. Here we cast those abstract results in a more concrete form. In particular, we show that, with some precisely defined physical entities, there is an intrinsic relationship between the Reynolds number, the number of degrees of freedom describing fluid flow, and the fractal dimension of the Navier-Stokes equations, i.e. the time asymptotic behavior of their solutions. Complicated nonlinear systems may sometimes behave in a counterintuitive manner; therefore, especially where numerical fluid flow simulations are concerned, one may be misled and arrive at erroneous conclusions. A recent example of such a pitfall was the observation that the chaotic behavior of the well-known Lorenz model of the Rayleigh-Benard convection disappears when the model is augmented by the addition of higher order modes [Franceschini and Tebaldi, 1981; Treve and Manley, 1982]. Thus here it is not only the quantitative nature of the approximation that is affected, but perhaps more radically, the qualitative nature of the approximation is radically changed.

What is then the degree of approximation needed to insure that a numerical solution of Navier-Stokes equations is at a minimum qualitatively correct? I.e, when a given approximation is steady, or periodic, or quasiperiodic, or aperiodic, under what conditions can one conclude that the exact solution the same property? Here we discuss very briefly certain important results relevant to three dimensional viscous incompressible flows. The corresponding rigorous analysis and a more extended

*US Department of Energy, Washington, DC 20545.

heuristic discussion are presented elsewhere [Constantin, Foias, Manley, and Temam, 1984a,b; Constantin, Foias and Temam, 1984], as are the results for the two dimensional case [Foias, Manley, Temam and Treve, 1983; Constantin and Foias, 1983].

In conventional turbulence theory [Landau and Lifshitz, 1959], one estimates the number N of degrees of freedom of a turbulent 3D-flow as

(1) $$N \sim (L_0/L_d)^3$$

where L_0 is the typical large scale length and L_d is the Kolmogorov dissipation length

(2) $$L_d = (\nu^3/\epsilon)^{1/4}$$

with ϵ the energy dissipation rate per unit mass, and ν the kinematic viscosity. Heretofore such an estimate has been based almost solely on otherwise unsubstantiated dimensional and order-of-magnitude arguments. Here we sketch out how starting with the Navier-Stokes equations one can deduce (1) in a rigorous way.

Before continuing, for the sake of notational simplicity, we define for any function, F, the upper bound on the time average of F

$$\overline{\langle F\rangle} = \lim_{t\to\infty} \sup\, (1/t)\int_0^t F(t')\, dt'$$

and the lower bound on that average

$$\underline{\langle F\rangle} = \lim_{t\to\infty} \inf\, (1/t)\int_0^t F(t')\, dt'$$

We consider the flow of an incompressible viscous fluid contained in a finite three dimensional region Ω, with a rigid boundary $\partial\Omega$, governed by the Navier-Stokes equations. Thus, the fluid velocity $\vec{v}(\vec{r},t)$ is determined by

(3a) $$\partial\vec{v}/\partial t + \vec{v}.\nabla\vec{v} = -\nabla p + \vec{f} + \nu\, \nabla^2\vec{v}$$

(3b) $$\nabla.\vec{v} = 0$$

(3c) $$\vec{v}|_{\partial\Omega} = 0$$

(3d) $$\vec{v}(\vec{r},0) = \vec{v}_0(\vec{r})$$

where $\vec{f}$ is the external force per unit mass, and p is the pressure divided by density. Alternatively, Ω can be a prism with sides L_0, L_1, and L_2, the boundary condition (3c) being replaced by periodicity conditions and the condition [Temam,1983]

(3e) $$\int_\Omega \vec{v}(\vec{r},t) = 0$$

There are two natural ways in which the intrinsically finite number of degrees of freedom of a three dimensional flow can be made manifest, namely in terms of determining modes and in terms of the fractal dimension of the attractors. The concept of determining modes is discussed below in greater detail. By attractors, we mean the time-asymptotic limits of the flow. Their fractal dimensions are addressed subsequently.

2. Determining Modes. Let $(\vec{w}_m)_{m=1}^{\infty}$ be a complete orthonormal set of three dimensional vector-valued eigenfunctions in Ω satisfying Stokes' equations,

$$\nabla^2 \vec{w}_m + \nabla g_m = -\lambda_m \vec{w}_m, \quad \nabla \cdot \vec{w}_m = 0$$

which, together with the appropriate boundary conditions, determine uniquely g_m. Here the index m stands for subscripts arranged so that λs form an increasing sequence, $0 < \lambda_1 \leqslant \lambda_2 \leqslant \ldots$ Now consider the eigenfunction expansion of $\vec{v}$,

$$\vec{v}(\vec{r},t) = \sum_{m=1}^{\infty} c_m(\vec{v})\vec{w}_m(\vec{r})$$

where $c_m(\vec{v})$, in general functions of time, are the expansion coefficients with respect to $\vec{v}$. Consider a finite set of modes, $c_j(\vec{v})\vec{w}_j$, ($j = 1, 2, \ldots, M$) of a solution $\vec{v}(\vec{r},t)$ of (3). Let there be any other solution $\vec{u} = \vec{u}(\vec{r},t)$, starting from different initial conditions. Now let M be so large that, if the differences between the expansion coefficients $c_m(\vec{v})$ and $c_m(\vec{u})$ vanish for long time, i.e. if

$$\lim_{t\to\infty} |c_j(\vec{v}) - c_j(\vec{u})| = 0, \quad (j = 1, 2, \ldots, M)$$

than in some sense $\vec{u}$ and $\vec{v}$ are equal to one another, or more formally, that

$$\lim_{t\to\infty} \int_{\Omega} |\vec{u}(\vec{r},t) - \vec{v}(\vec{r},t)|^2 = 0.$$

Then such a set of M modes is said to be determining [Foias et al, 1983].

From the practical point of view, the importance of determining modes lies in that in many respects, such as stability, periodicity, etc., the behavior of the approximation to v(r,t) consisting only of those modes is the same as that of the true solution. However, no such assurance obtains if the approximation is of lesser order than that based on the determining modes.

As is well known, it is not yet certain that regular solutions of (3) exist for all times. However, following common practice we assume that in some sense most of such solutions are regular, and in particular that the vorticity is bounded. We define the maximum dissipation rate, ϵ, as

$$(4) \quad \epsilon = \nu \overline{\langle \sup_{\vec{r}} |\nabla \vec{u}(\vec{r},t)|^2 \rangle}$$

where $|\nabla\vec{u}(\vec{r},t)|^2 = \sum_{i,j=1}^{3} |\partial u_i/\partial x_j|^2$. For the present purposes the Kolmogorov length will be defined by (2) with the above value of ϵ. Further, we let $L_o = \lambda_1^{-1/2}$.

We demonstrate now the validity of (1), or more precisely we show that a sufficient number, M, of the determining modes of $\vec{u}$, is approximately equal to $k_1 N$. Here and in the sequel $k, k_1, k_2, \ldots$ denote dimensionless absolute constants, typically of the order of unity, depending at most on the shape of the flow field, but not on its size.

On taking the difference between the equations (3a) for $\vec{u}$ and $\vec{v}$, multiplying the result by $C_j(\vec{u},\vec{v})\vec{w}_j$, where $C_j(\vec{u},\vec{v}) \equiv c_j(\vec{u}) - c_j(\vec{v})$, summing over j from M + 1 to ∞, and integrating over the volume, we obtain

$$d/dt \sum_{j=M+1}^{\infty} [C_j(\vec{u},\vec{v})]^2 + 2(\nu\lambda_{M+1} - \sup_{\vec{r}} |\nabla\vec{u}|) \sum_{j=M+1}^{\infty} [C_j(\vec{u},\vec{v})]^2$$
$$< k\sum_{i=1}^{M} C_i(\vec{u},\vec{v}) \times (\text{other terms})$$

which vanishes as $t \to \infty$, because by hypothesis for $1<i<M$, in that limit $C_i(\vec{u},\vec{v}) \to 0$.

Hence we easily infer that $[c_1(\vec{u})\vec{w}_1, \ldots, c_M(\vec{u})\vec{w}_M]$ is a determining set for $\vec{u}$, provided that

$$(5) \qquad \nu\lambda_{M+1} - \overline{\langle \sup_{\vec{r}} |\nabla\vec{u}| \rangle} > 0$$

But it is known that in three dimensions $\lambda_{M+1} > k_2\lambda_1 M^{2/3}$ [Morse and Feshbach, 1953; Metivier, 1978], so that (5) holds if

$$\overline{\langle \sup |\nabla\vec{u}| \rangle} < k_2 M^{2/3} \nu\lambda_1$$

or more pertinently, if

$$\epsilon^{1/2} \leq k_2 M^{2/3} \nu^{2/3} \lambda_1 = k_2 M^{2/3} \nu^{3/2} / L_o^2$$

That is, it suffices that the number of modes $N \geq M \sim k_2^{-3/2} (L_o/L_d)^3$, or to within a constant, we recover the estimate (1). A novel conclusion to be drawn from this result is that the conventional estimate (1) is really an upper bound on the number of modes needed to describe 3D turbulent flow.

The derivation leading up to (5) justifies the common assumption about the physical basis for (1). We see that indeed when (5) is satisfied, the length scale given by λ_{M+1} is sufficiently small to insure that the energy delivered to the higher order modes by shear stress at the rate ~ $|\nabla u|$ is effectively damped by molecular viscosity.

3. Variation Equations and Dimensions of Attractors. A natural way to determine the dimension of a space in which time asymptotic trajectories are imbedded is to test in some manner the neighborhood of a given

trajectory. In particular, such a test may consist of examining how a small volume element evolves along that trajectory: for, if we find that with a given assigned spatial dimensionality, say D, the small D-volume is shrinking indefinitely, we can conclude that, for t--> , the trajectories cannot fill a D-dimensional volume.

Consider then the so-called variation equations associated with a given system of n differential equations [Pars, 1965]:

$$\dot{x} = f(x) \qquad x = \{x_1, x_2, \ldots, x_n\}$$

Let F(t) be a matrix with elements $\partial f_i[x(t)]/\partial x_j$, with $x_j(t)$ assumed to be known functions of time. Then the variation equations are

$$(6) \quad \dot{z} = F(t)z \qquad z = \{z_1, z_2, \ldots, z_n\}$$

Now consider an n-dimensional infinitesimal volume element

$$Y = |y_1 \wedge y_2 \wedge \ldots \wedge y_n|$$

moving along the trajectory prescribed by x(t), where y_i, (i=1,2,...,n) are some solutions of (6), and ^ signifies the outer vector product. Its evolution is evidently

$$(7) \quad \dot{Y}/Y = \sum_{i=1}^{n} \partial f_i[x(t)]/\partial x_i \equiv \text{Tr } F(t).$$

Therefore if Tr F < 0, the n-volume is shrinking. Moreover, it can be shown then that the dimension of the region in which x(t) resides asymptotically as $t \to \infty$, is less than n. A familiar example of such behavior is the well-known Lorenz system, which is of third order, but whose asymptotic trajectory lies on a complicated set - the Lorenz attractor - which is in fact of lesser than three dimensions [Lanford, 1976]. Of course, less exotic examples are limit cycles (one dimension), and point attractors (zero dimension). We now recall here that it has been conjectured that the dimension of an attractor is intimately related to the so-called Lyapounov characteristic numbers, or Lyapounov exponents [Kaplan and Yorke, 1979; Russell, Hanson and Ott, 1980]. This enormously useful relationship has now been proved rigorously [Constantin and Foias, 1983; Constantin et al, 1984; Babin and Vishik, 1983].

Consider the volume element $Y_m = |y_1 \wedge y_2 \wedge \ldots \wedge y_m|$, where y_i are $m \leq n$ solutions of (6). Further, initially let $y_i(0)$ be an orthonormal set of vectors, i.e., initially, to within an appropriate constant depending only on m, Y_m is the volume of an m-dimensional sphere. Then (6) serves to describe the deformation of that sphere into an m-dimensional ellipsoid, and its subsequent evolution. At any time the m-volume of that ellipsoid is proportional to the product of its m semiaxes. Let $\alpha_1(t), \alpha_2(t), \ldots, \alpha_m(t)$ denote the time dependent semiaxes of the ellipsoid. Then

$$(8) \quad \int_0^t \text{Tr } F_m(t')\, dt' = \ln \omega_m(t)$$

where $\omega_m(t) = \alpha_1(t)\alpha_2(t)\ldots\alpha_m(t)$, and Tr $F_m(t)$ denotes the trace of the matrix F(t) restricted to the space spanned by the principal axes of the ellipsoid. Now we allow t to become so large that it is reasonably certain that u(t) is in the attractor, or very close to it. Assume that the limit

$$(9) \quad \mu_i = \lim_{t\to\infty} \sup\, [\ln \alpha_i(t)]/t$$

exists, then μ_i is the Lyapounov exponent associated with $\alpha_i(t)$ [Osedelec, 1968]. Note that in the most general case $[\ln \alpha_i(t)]/t$ may not tend to a well defined limit. However as long as the α_i's are bounded (9) is meaningful. It follows then from (8) that as $t \to \infty$, and the m-volume moves along a trajectory in the attractor of a system of n differential equations, the time integral of the trace of F_m satisfies

$$(10) \quad (1/t)\int_0^t \mathrm{Tr}\, F_m(t')\, dt' \sim \sum_{i=1}^{m} \mu_i$$

On comparing (8) with (10) we see that we have thus related the time evolution of elements of different dimensions in the n-dimensional phase space to the sum of the appropriate Lyapounov exponents. It has been shown elsewhere that for a wide class of dynamical systems, including differential equations of practical interest, the required limits, and hence Lyapounov exponents exist for most trajectories [Osedelec, 1968]. When $\sum_{i=1}^{m} \mu_i$ is consistently negative, one can prove that the dimension of the attractor is less than m.

For certain technical reasons [Constantin et al, 1984], we introduce the so-called uniform Lyapounov exponents, μ_i, defined somewhat differently than those in (9). Specifically, they are determined iteratively as

$$(11) \quad \mu_i = \lim_{t\to\infty} \{\ln[\overline{\omega}_i(t)/\overline{\omega}_{i-1}(t)]\}/t$$

where $\overline{\omega}_i(t) = \sup \omega_i(t)$, with the supremum taken over all trajectories in the attractor.

Now recall a generalization of the concept of dimension, the so-called fractal dimension, $d_M(X)$, of a set X: it is based on the number of small volumes needed to fill a region of space [Mandelbrot, 1977]. More precisely the fractal dimension, $d_M(X)$, of an object X is defined as

$$d_M(X) = \lim_{\epsilon\to 0} \sup[\ln n_x(\epsilon)]/\ln(1/\epsilon)$$

where $n_x(\epsilon)$ is the minimum number of balls of radii $\leq \epsilon$ needed to fill, or cover X. It is clear that $n_x(\epsilon)$ depends on the volume of the "balls", or in the present case on $\omega_m(t)$. Indeed, a laborious analysis [Constantin and Foias, 1983; Constantin et al, 1984] shows that

$$(12) \quad d_M(X) \leq \max_{1\leq \ell\leq m} \left(\ell + \sum_{i=1}^{\ell} \mu_i / |\bar{\mu}_{m+1}|\right) \leq \max_{1\leq \ell\leq m} \left[\ell + (m+1)\sum_{i=1}^{\ell} \mu_i \Big/ \left|\sum_{i=1}^{m+1} \mu_i\right|\right]$$

Here $\bar{\mu}_i = \lim_{t\to\infty} \sup[\ln \bar{\alpha}_i(t)]/t$, and $\bar{\alpha}_i = \sup \alpha_i(t)$, with the supremum taken again over all the trajectories in the attractor. For practical application the rightmost member of (12) is easiest to evaluate. The best value for n is the first integer such that $\sum_{i=1}^{m} \mu_i \geq 0$, while $\sum_{i=1}^{m+1} \mu_i < 0$.

We now extend these elementary ideas to the attractor of the Navier-Stokes equations.

4. Fractal Dimension of the Navier-Stokes Attractor. In order to apply the results of the preceding section, it suffices, without any loss of generality, to restrict oneself to the solenoidal (divergence-free) portion of (3). Then the equation of evolution corresponding to (3) is of the form [Temam, 1983]

$$(13) \qquad d\vec{v}/dt + \nu A\vec{v} + B(\vec{v},\vec{v}) = \vec{f}$$

where we have assumed that the force, f, is a divergence-free vector. Here $-A$ and $B(\vec{p},\vec{q})$ are the divergence-free parts of the Laplacian (∇^2) and $\vec{p}.\nabla\vec{q}$, respectively, satisfying their appropriate boundary conditions. Next we linearize (13) about some known solution $v_o(\vec{r},t)$, by taking the Frechet derivative of (13). This leads to a variation-like equation for a small departure, z, from the trajectory v ,

$$(14) \qquad d\vec{z}/dt + \nu A\vec{z} + B(\vec{v}_o,\vec{z}) + B(\vec{z},\vec{v}_o) = 0$$

Note that because of the orthogonality of $B(\vec{p},\vec{q})$ with respect to $\vec{q}$, there is no contribution to the trace from $B(\vec{v}_o,\vec{z})$. We define $Q_m(\vec{U}_1,\vec{U}_2,\ldots,\vec{U}_m)$, where $\vec{U}_i$ are solutions of (14), to be the orthogonal projector onto the m-dimensional space (imbedded in an infinite dimensional phase (function) space). For notational simplicity we designate the finite, mth rank operators $A_m = AQ_m$, and $B_m = BQ_m$, with the corresponding traces Tr A_m and Tr B_m. Further, we let $\mathcal{R}_m = \nu A_m + B_m$. Then we have

$$d \ln(|\vec{U}_1 \wedge \ldots \wedge \vec{U}_m|^2)/dt = -2\mathrm{Tr}\,\mathcal{R}_m(\vec{v}_o)$$

It is possible to prove that there exists a number N_o, such that if $m \geq N_o$, a small m-dimensional volume evolving along the solution of (13) shrinks indefinitely , as $t \to \infty$ [Constantin and Foias, 1983; Constantin et al, 1984]. The key result is that $N_o \sim N$, as in (1). The proof is very technical and elaborate, and it will not be reproduced here. However, in order to give the reader some idea of what is involved, the two key steps are discussed very briefly below.

In the first step, let X be any attractor such that for all solutions of (13) as $t \to \infty$ the enstrophy is bounded from above, i.e.,

$$\sup_{\vec{v}\in X} \int_\Omega |\nabla\vec{v}|^2 < \infty$$

Here again the volume, ω_m, of the m-dimensional ellipsoid evolving along the solution of (13) is

$$\omega_m = \exp(-\int_0^t \mathrm{Tr}\, \mathcal{A}_m dt')$$

where we have assumed that initially that volume was an m-dimensional unit sphere. Now, for any m = 1, 2,... define the smallest possible time averaged value of the trace of $\mathcal{A}_m$, q_m, for all possible projectors Q_m, for a given m:

$$(15) \quad q_m = \inf_{\vec{v}} \langle \inf_{Q_m} \mathrm{Tr}\, \mathcal{A}_m \rangle$$

where $\vec{v}$ runs over all solutions of (13) with the appropriate boundary condition, and such that the initial conditions $\vec{v}(\vec{r},0)$ are already in the attractor $\mathbf{X}$. Evidently, q_m is a bound on the sum of the m largest uniform Lyapounov exponents. Then for $q_m > 0$, the analog of (12) in infinite dimensional space yields for the fractal dimension

$$(16) \quad d_M(\mathbf{X}) < m[1 + \max_{1 \le \ell \le m} (-q_\ell)/q_m]$$

In the second step, we determine q_ℓ and q_m, followed by the demonstration that $d_M(Y) \sim N$ of (1). First, let the upper bound on the energy dissipation rate be given by

$$(17) \quad \epsilon = \overline{\langle \sup_{\vec{v}} \sup_{\vec{r}} |\nabla \vec{v}(\vec{r},t)| \rangle} < \infty$$

where, as above, $\vec{v}$ runs over all solutions of (13) which are in $\mathbf{X}$. We define now L_d by (2) and (17) instead of (2) and (4).

It can be shown that on using (17)

$$(18) \quad |\mathrm{Tr}\, B_m(.,\vec{v}(\vec{r},t))| < m \sup|\nabla \vec{v}(\vec{r},t)| \leq m(\epsilon/\nu)^{1/2}$$

and that

$$(19) \quad \mathrm{Tr}\, A_m > \lambda_1 + \lambda_2 + \ldots + \lambda_m > k_2 \lambda_1 (1 + 2^{2/3} + \ldots + m^{2/3})$$

Therefore it follows on substituting (18-19) in (15) that for any positive integer, m,

$$q_m > m\nu\lambda_1(3k_2 m^{2/3}/5 - N^{2/3})$$

From elementary considerations it follows that for $k_2 m^{2/3} > 2.5N^{2/3}$

$$\max -q_\ell \leq 0.4\nu\lambda_1 N^{5/3}/k_2^{3/2}$$

On substituting in (16) we obtain finally,

$$(20) \quad d_M(\mathbf{X}) < 4.75N/k_2^{3/2}$$

Therefore, apart from a constant of order unity, we find that the bound on the dimension of the attractor of the Navier Stokes equation is identical with the estimate of the number of modes sufficient for the description of the time asymptotic behavior of the solutions to that equation.

Similarly, on using methods akin to those leading to (18-19), it is possible to show from (15) that

$$q_m > \nu\lambda_1 m(3k_2 m^{2/3} - R^2)/2.$$

where the Reynolds number, R, is defined consistently with the preceding as

$$R = \overline{\langle \sup_{\vec{v},\vec{r}} |\vec{v}(\vec{r},t)|^2 \rangle^{1/2}} / \nu\lambda_1^{1/2}$$

From this it follows in turn that the use of arguments leading up to (20) results in

$$(21) \quad d_M(X) \sim R^3 .$$

This is a somewhat pessimistic estimate, being much higher than the conventionally accepted N ~ $R^{9/4}$ [Landau and Lifshitz, 1959]. However it must be noted that in arriving at (21), we have made no use of any a priori knowledge of the spectrum of the flow. The more conventional estimate is based on the assumption that the spectrum of the energy density in the flow is the Kolmogorov spectrum.

5. Concluding Remarks. First, we have been able to relate the bounds on the degrees of freedom and fractal dimension of the Navier-Stokes attractor to an appropriately defined Reynolds number. We have remarked on the conventional measure of the number of degrees of freedom, $R^{9/4}$, as being a result of a priori knowledge of the spectrum of homogeneous, isotropic turbulence. Obviously this conventional estimate is inapplicable at low and medium sized Reynolds numbers. A more conservative estimate, independent of the knowledge of the spectrum, as suggested here, varies as R^3.

Secondly, an overriding conclusion emerging from the work reported here is that in every sense the Navier-Stokes equation is a closed system. That is, it is determined by a finite number of degrees of freedom. Thus, were one to carry a sufficient number of terms in an approximate solution, the terms beyond become irrelevant.

Finally, the key to obtaining the results presented in this paper lies in the ability to estimate the values of certain integrals, i.e., certain norms. The limitation on those results, their being only sufficient bounds, but not necessary and sufficient, rests on the limits to our present ability to make such estimates.

REFERENCES

BABIN, A. V., and VISHIK, M. I. 1983 Attractors of Partial Differential Equations of Evolution and Estimates of their Dimensions. Usp. Math. Nauk, 38, 4 (232), 133. (In Russian).

CONSTANTIN, P., and FOIAS, C. 1983 Global Lyapounov Exponents, Kaplan-Yorke Formulas and the Dimension of the Attractor for 2D Navier-Stokes Equations. Comm. Pure and Applied Math. (To appear).

CONSTANTIN, P., FOIAS, C., MANLEY, O. P., and TEMAM, R. 1984a C. R. Acad. Sc., 297 Serie I, 599.

CONSTANTIN, P., FOIAS, C., MANLEY, O. P., and TEMAM, R. 1984b Determining Modes and Fractal Dimension of Turbulent Flows, J. Fluid Mech., (Submitted for publication).

CONSTANTIN, P., FOIAS, C., and TEMAM, R. 1984 Attractors Representing Turbulent Flows. Memoirs of the American Mathematical Society. (To appear)

FOIAS, C., MANLEY, O. P., TEMAM, R., TREVE, Y. M. 1983 Asymptotic Analysis of the Navier-Stokes equations, Physica 9D, 157.

FRANCESCHINI, V., and TEBALDI, C. 1981 J. Stat. Phys. 25, 397.

KAPLAN, J., and YORKE, J. A. 1979 Chaotic Behavior of Multidimensional Difference Equations. Functional Differential Equations and Approximation of Fixed Points, H. O. Peitgen and H. O. Walther, Editors, Lecture Notes in Mathematics, vol. 730. Springer.

LANDAU, L. D., and LIFSHITZ, E. M. 1959 Fluid Dynamics, Addison-Wesley.

LANFORD III, O. E. 1976 Qualitative and Statistical Theory of Dissipative Systems, CIME School of Statistical Mechanics.

MANDELBROT, B. 1977 Fractals. Freeman.

METIVIER, G. 1978 J. Math. Pures Appl. 57, 133.

MORSE, P. M., and FESHBACH, H. 1953 Methods of Theoretical Physics. McGraw-Hill.

OSEDELEC, V. I. 1968 A Multiplicative Theorem. Lyapounov Characteristic Numbers for Dynamical Systems. Trans. Moscow Math. Soc. 19, 197.

PARS, L. A. 1965 A Treatise on Analytical Dynamics, Heineman.

RUSSELL, D. A., HANSON, J. B., and OTT, E. 1980 Phys. Rev. Lett. 45, 1175.

TEMAM, R. 1983 *Navier-Stokes Equations and Nonlinear Functional Analysis.* NSF/CBMS Regional Conference Series in Applied Mathematics, SIAM.

TREVE, Y. M., and MANLEY, O. P. 1982 *Physica* *4D*, 319.

CHAOTIC OSCILLATIONS IN MECHANICAL SYSTEMS

EARL H. DOWELL* AND CHRISTOPHE PIERRE*

I. Introduction. In recent years a great deal of progress has been made in the study of chaos in nonlinear mechanics. A recent, very readable and informative survey of much of the work is contained in the paper by Holmes and Moon [1]. Broadly speaking most of the dynamical systems studied to date can be categorized as either (1) forced response of stable (with respect to infinitesimal disturbances) dynamical systems or (2) self excited oscillations of autonomous systems. Representative of the first category is the study of the forced response of Duffings Equation with a negative linear stiffness and a positive cubic stiffness. This, for example, models the behavior of a buckled beam or plate under transverse dynamic excitation [2]. Representative of the second category is the set of Lorenz equations [3] or the panel flutter equations [4,5].

A fundamental question is what mechanisms in these two categories of systems lead to chaotic oscillations. Are the mechanisms similar or are they distinct? Recent work by the present author and others suggests that one possible distinction is that in the first category the chaotic oscillations arise as a result of instability of the system with respect to large finite disturbances while in the second category the chaos results from instability with respect to infinitesimal disturbances. This brings out the importance of the initial value problem in understanding category one chaos and also the importance of being able to calculate efficiently and systematically the equilibria whose stability with respect to finite or infinitesimal disturbances is to be examined. Also of importance is the development of ways in which to interpret the results of computer simulations or experiments on dynamical systems undergoing chaotic motions. These topics are discussed in the following.

*Duke University, Durham, NC 27706

II. Forced Response of a Buckled Beam (Category One)

• Response to an Initial Impulse (One Mode Model)

Recall that the single mode model [1,2] takes the form of a non-linear ordinary differential equation for the modal amplitude, $A(\tau)$,

$$\ddot{A} + \gamma\dot{A} - \frac{A}{2}(1 - A^2) = F(\tau) \tag{1}$$

with the initial conditions

$$A(\tau = 0) = A_0 \qquad \dot{A}(\tau = 0) = \dot{A}_0 \tag{2}$$

τ is a non-dimensional time, γ is a damping coefficient and F is the excitation force. Preliminary calculations have been made for the initial value or impulse problem, i.e. $F(\tau) \equiv 0$. Specifically, we shall first consider $A_0 \equiv 1$, and $\dot{A}_0 \neq 0$. Note that $A = 0, +1, -1$ are static equilibrium solutions for (1) with $F \equiv 0$. Further, as is well known, the static equilibria solutions, $A = \pm 1$ are stable with respect to infinitesimal disturbances, but the static equilibrium solution, $A \equiv 0$, is unstable with respect to infinitesimal disturbances.

The question arises whether the static equilibria, $A = \pm 1$, are unstable with respect to finite disturbances. On physical grounds, one might expect such finite disturbances could drive the solution from one static equilibrium, say $A = +1$, to the other, $A = -1$. This, in fact, occurs. What is, perhaps, more surprising is that transition from one static equilibria to the other has a very complicated and essentially statistical dependence on the magnitude of the finite disturbances. This (extreme) sensitivity of the steady state ($\tau \to \infty$) solution, i.e. $A = +1$ or -1, on the initial conditions, A_0 and $\dot{A}_0$ is one form of what has been termed "chaos".

One finds the following:

For $F(\tau) \equiv 0$ and $A_0 = 1$, for various $\dot{A}_0$, the final modal amplitude $A(\tau \to \infty)$ may be either +1 or -1. Specifically for $0 < \dot{A}_0 < .521799$ one finds that $A(\tau \to \infty) = +1$. A typical phase plane trajectory is shown in Fig. 1 for $\dot{A}_0 = .52$. Note that $A > 0$ for all τ.

As $.521799 < \dot{A}_0 < .5572$, $A(\tau \to \infty) \to -1$. A phase plane trajectory for $\dot{A}_0 = .54$ is shown in Fig. 2. Note there is a single crossing of the $A = 0$ axis.

As $.5572 < \dot{A}_0 < .5952$, $A(\tau \to \infty) \to +1$. A phase plane trajectory for $\dot{A}_0 = .56$ is shown in Fig. 3. Note there are now two crossings of the $A = 0$ axis.

The pattern is now clear. As $\dot{A}_0$ increases, there are alternating intervals of width, $\Delta\dot{A}_0$, which lead to a final modal amplitude of +1 or -1. One can associate an integer index, N, with each interval which is equal to the number of crossings of the A = 0 axis. It is of special interest to consider how $\Delta\dot{A}_0$ varies with N. For technical reasons it is preferable to consider the relative interval width, $\Delta\dot{A}_0/\dot{A}_0$.

In Fig. 4a $\Delta\dot{A}_0/\dot{A}_0$ is plotted vs. N. For large N, $\Delta\dot{A}_0/\dot{A}_0 \to 0$. The rate at which this asymptote is approached is at nearly 1/N. In Fig. 4b a similar plot is shown for $\Delta A_0/A_0$ vs N for $A_0 \equiv 0$. Here the asymptote appears to be approached at a rate of very nearly 1/N.

To summarize, the ranges of initial velocity, $\dot{A}_0$, alternate which lead to one final modal amplitude or the other. Moreover the relative width of each of these ranges progressively diminishes as $\dot{A}_0$ increases. Thus, for sufficiently high $\dot{A}_0$, and given any uncertainty in the precise value of $\dot{A}_0$, the final modal amplitude becomes unpredictable. A similar conclusion applies to initial displacement, A_0.

Further calculations are in progress to understand better this fascinating result. Specifically a variety of initial conditions, A_0 and $\dot{A}_0$, will be studied to generalize the one dimensional result of Fig. 4 to two dimensions in terms of <u>both</u> A_0 <u>and</u> $\dot{A}_0$. Heuristically, it is expected such a two-dimensional representation may take the form (for small damping, $\gamma \to o$) of contours where the sum of initial kinetic plus potential energy is a constant, i.e.

$$T_0 + V_0 = \frac{1}{2}\dot{A}_0^2 - \frac{A_0^2}{4}\left(1 - \frac{A_0^2}{2}\right) \tag{3}$$

This will be verified or refuted by undertaking systematic calculations.

The model and associated calculations will also be extended to include additional beam modes to see under what circumstances such modes may be qualitatively or quantitatively important.

- Forced Response

Once a thorough understanding of the beam response to an initial impulse is obtained, the forced response, $F(\tau) \neq 0$, will be reconsidered. Specifically,

- as needed, multimodes will be included in the model
- a simple criterion will be sought to characterize under what conditions chaotic oscillations will occur

- a quantitative comparison between theory (calculations) and (physical) experiment will be made

III. Self-Excited, Chaotic Oscillations of an Autonomous System (Category Two)

Self-excited chaotic oscillations appear to arise from a somewhat different mechanism than that associated with forced oscillations. Specifically in the case of the Lorenz equations [3], such chaotic oscillations appear to be the result of all (static or dynamic) equilibrium solutions becoming unstable with respect to infinitesimal disturbances. It is planned that

- a systematic study of the static and dynamic equilibria will be made using recently developed techniques such as the incremental harmonic balance method, [6,7]

- followed by a systematic study of the stability of such equilibria with respect to infinitesimal disturbances using conventional or Floquet stability theory

Such studies should further our understanding of the mechanism by which chaotic oscillations arise.

Here some of the important presently known characteristics of the solutions of the Lorenz equations are summarized. Also a new, recently developed [5] method is described for displaying and interpreting the results of computer simulations of these equations.

The Lorenz equations are:

$$\dot{x} = \sigma(y - x) \tag{4}$$

$$\dot{y} = -xz + rx - y \tag{5}$$

$$\dot{z} = xy - bz \tag{6}$$

Much of our present understanding of these equations and their solutions is summarized in the excellent book by Sparrow [8]. It is clear from (4)-(6) that, if x,y,z is a solution, then so is $-x,-y,z$.

- Static Equilibrium Solutions

It is easily shown that for $r < 1$, only a single static equilibrium solution exists, namely

$$x_0 = y_0 = z_0 = 0 \tag{7}$$

For $r > 1$, three static equilibrium solutions exist, one is th given by (7) and the two others are

$$x_0 = y_0 = \pm b^{1/2}(r-1)^{1/2}$$

(8) $$z_0 = r - 1$$

- Stability of Static Equilibria with Respect to Infinitesimal Disturbances

In the usual way, one assumes that:

$$x = x_0 + \hat{x}$$

(9) $$y = y_0 + \hat{y}$$

$$z = z_0 + \hat{z}$$

and substitutes (9) into (4)-(6). Linearizing the result in $\hat{x}$, $\hat{y}$, $\hat{z}$ and using (7) or (8), one can determine the stability or instability of the static equilibria, (7) or (8), by standard eigenvalue methods.

It is found that the equilibrium described by (7) is stable with respect to infinitesimal disturbances for $0 < r < 1$ and unstable for $r > 1$. The equilibria described by (8) do not exist for $0 < r < 1$, of course. For $r > 1$ they are stable until some larger $r = r^*$ where r^* depends upon σ and b. For the commonly studied values of $\sigma = 10$ and $b = 8/3$, $r^* = 24.74$.

For $r > r^*$, where all three static equilibria are unstable with respect to infinitesimal disturbances, chaos ensues. One might reasonably ask, however, whether there are any stable (large amplitude) dynamic equilibria (limit cycles) for $r > r^*$. Apparently the answer is no. The evidence for this last statement is twofold. First of all, numerical solutions obtained by digital simulation show chaotic (random-like) motions and no evidence of periodic, limit cycle motions. Secondly Sparrow has used the so-called "shooting" technique to obtain solutions for <u>unstable</u> periodic, limit cycle motions. These appear near $r = 13.926$ with infinite period in what Sparrow calls a "homoclinic explosion". The period decreases, as does the amplitude of this unstable limit cycle motion, as r increases. At $r = 24.74$, the amplitude of this unstable limit cycle appears to go to zero.

Of course, these unstable limit cycle motions would never be seen in a physical experiment or a digital simulation (numerical experiment). Nevertheless the study of their birth and death does provide additional understanding and insight into the transition to chaos as r increases.

• Boundedness of Large Amplitude Motion

When all (static or dynamic) equilibria become unstable with respect to infinitesimal disturbances there is the question as to whether the motion will remain bounded or not. Lorenz and Sparrow have shown that the motion will remain bounded by using Liapunov function methods. Here an alternative and simpler approach is used.

For sufficiently large motions, equations (4)-(6) simplify to

(4a) $$\dot{x} = \sigma(y - x)$$

(5a) $$\dot{y} = - xz$$

(6a) $$\dot{z} = xy$$

From (5a) and (6a), one concludes that

(10) $$y\dot{y} + z\dot{z} = 0 \quad \text{or } y^2 + z^2 = C, \text{ a constant}$$

From (10), one sees that if y,z are bounded at t = 0, then y,z are bounded for all time. Moreover from (4a),

if $x < y$, then $\dot{x} > 0$ and x increases

and if $x > y$, then $\dot{x} < 0$ and x decreases

Thus x is bounded and x is of the same order of magnitude as y.

• Some Representative Results

Several results will be shown for

• phase plane trajectories

• Poincare maps

• probability density of an event

In Fig. 5 a (two-dimensional cut of the) phase plane trajectory is shown for $r = 25$, $\sigma = 10$ and $b = 8/3$ in terms of z vs y. Each point is at a discrete instant of time. The phase plane trajectories spiral in a chaotic manner about the two non-trivial (and unstable) static equilibrium points given by (8).

In Fig. 6 a Poincare map is shown in terms of x vs y. An <u>event</u> is defined to occur when, say, $z = z_0 \equiv r - 1$. At those instants of time when an event occurs (which are separated in general by <u>random</u>

intervals of time), if x is plotted versus y, then a Poincare (or return, i.e. z returns to its chosen value) map is generated. As can be seen, the regions of possible values of x and y when an event occurs are relatively small. Such a map may be interpreted as displaying a conditional probability. That is, given the occurrence of an event, $z = z_0 \equiv r - 1$, and a value of say x, it answers the question what is the probable value of y.

Finally, in Fig. 7 the temporal statistics of an event are shown. Using the same parameters as before and the same definition of an event, the probability density of the time interval between two successive events is displayed. On the vertical axis the relative number of events which occur a certain time interval apart is plotted and on the horizontal is plotted the time interval itself. As can be seen there is a well defined most probable time interval between two successive events. This gives an important insight into the characteristic time scale of the chaotic oscillation. For more details on the temporal statistics of an event, see Ref. 5.

CONCLUDING REMARKS

As Sparrow [8] has noted, "If finite-dimensional equations are to be used to model infinite dimensional systems, it seems important that similar behaviors should be observed in low dimensional systems of differing dimension". In the case of both the buckled beam and the Lorenz equations models, there is a question of convergence or closure of the model equations. That is, if additional terms are retained in the spatial modal (generalized Fourier) series which are used to derive the model ordinary differential equations from the original partial differential equations (either beam theory or the Navier-Stokes equations), does the series expansion converge and, if so, how rapidly? Presently the answer to this question is not known. It is likely that the beam model converges rapidly while there is considerable doubt about the Lorenz model as Manley [9] and others have discussed. Interestingly the panel flutter model discussed in Ref. 4 and 5 is known to have good convergence properties.

ACKNOWLEDGEMENT

This work was supported, in part, by NSF Grant MEA-8315193 and AFOSR Grant 83-0346. Drs. Elbert Marsh and Anthony Amos, respectively, are the technical monitors. The authors would like to thank Mr. Michael D'Antonio for his help with the computations.

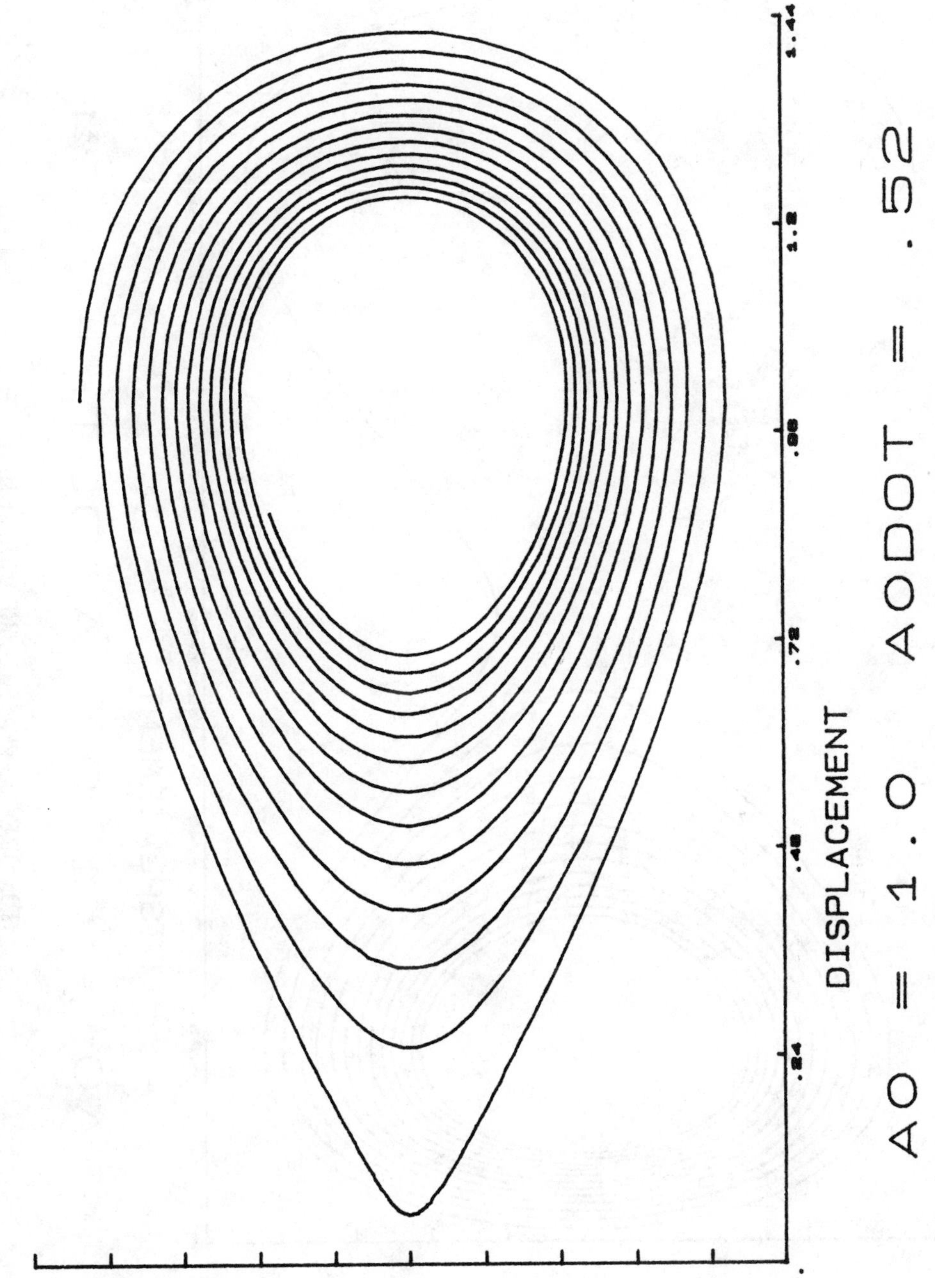

FIGURE 1. PHASE PLANE TRAJECTORY

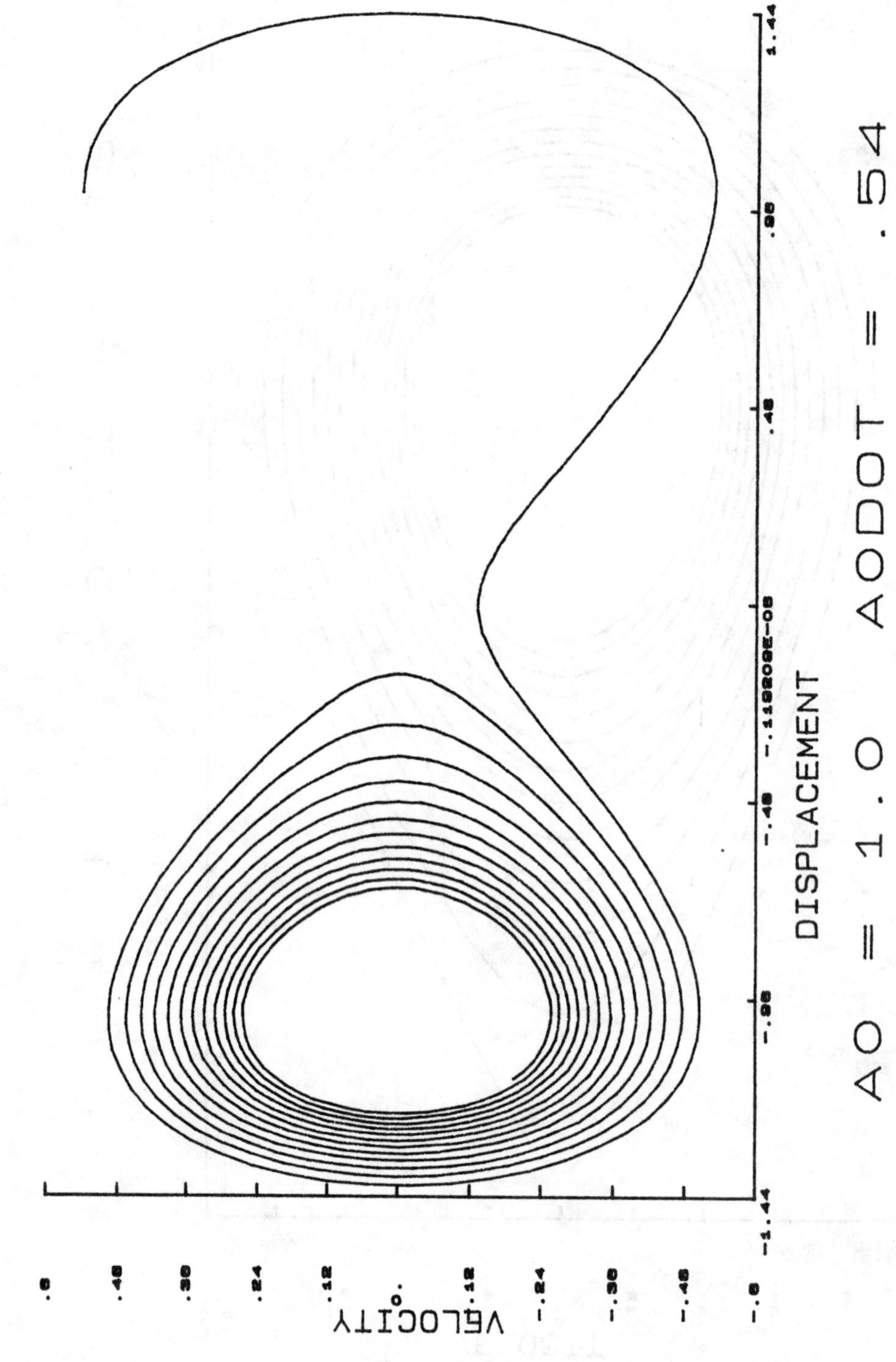

FIGURE 2. PHASE PLANE TRAJECTORY

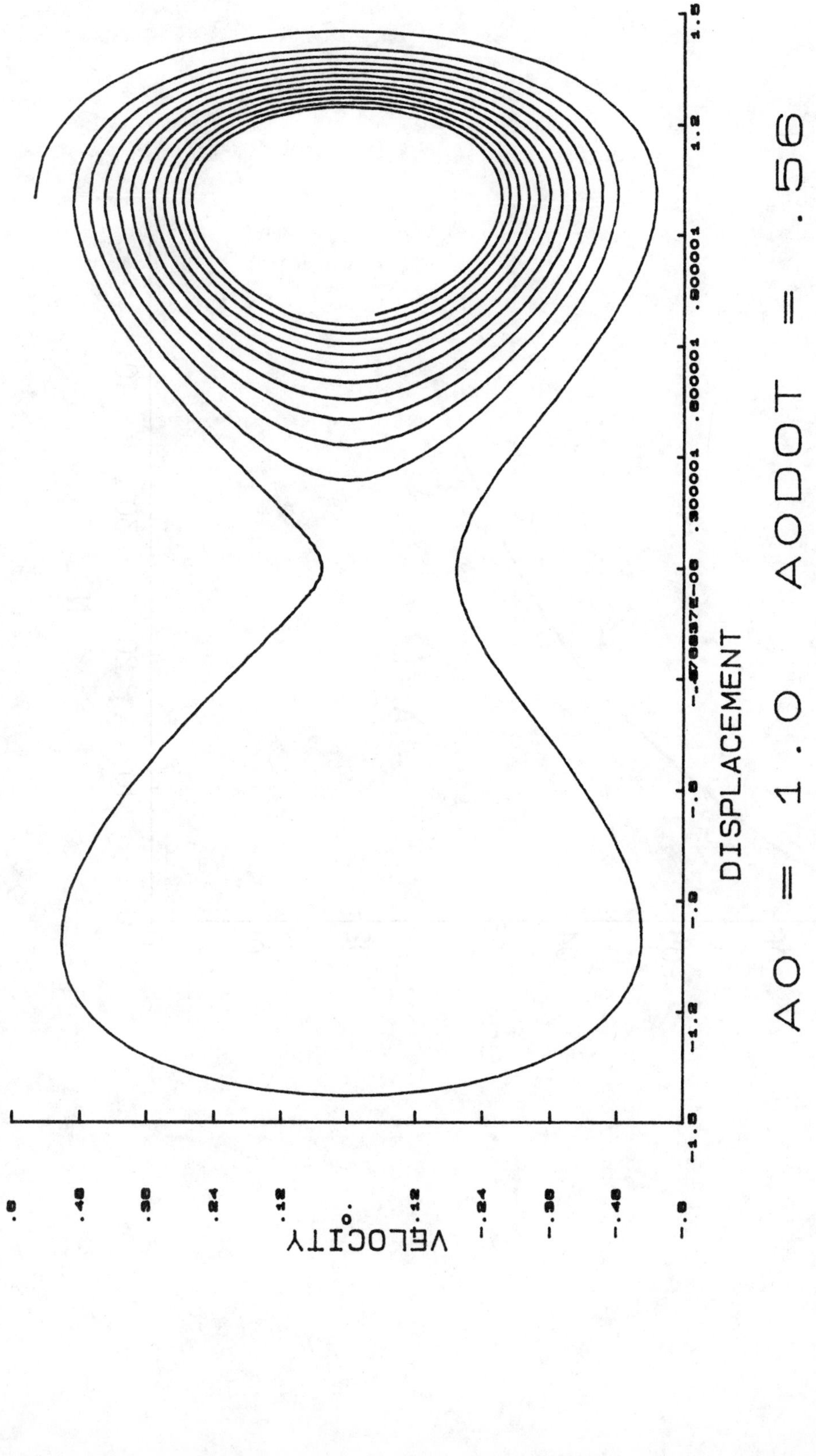

FIGURE 3. PHASE PLANE TRAJECTORY

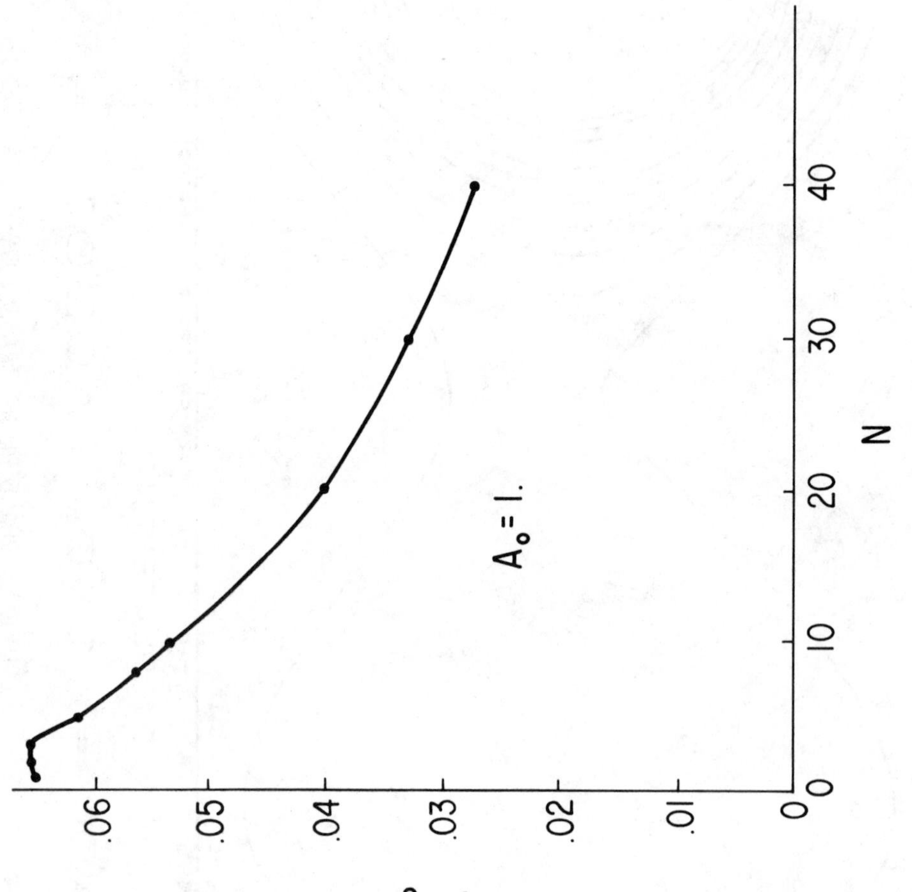

FIGURE 4a. RELATIVE VELOCITY RANGE vs NUMBER OF CROSSINGS

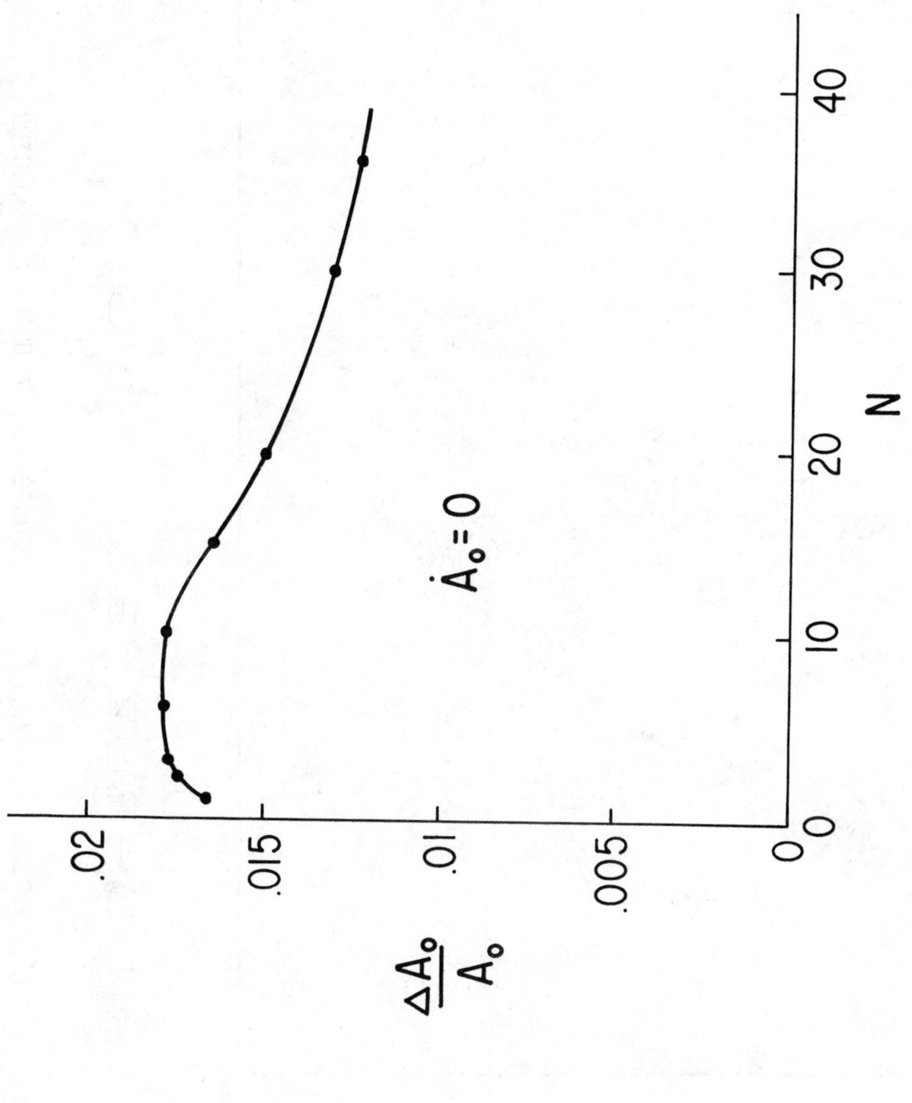

FIGURE 4b. RELATIVE DISPLACEMENT RANGE vs NUMBER OF CROSSINGS

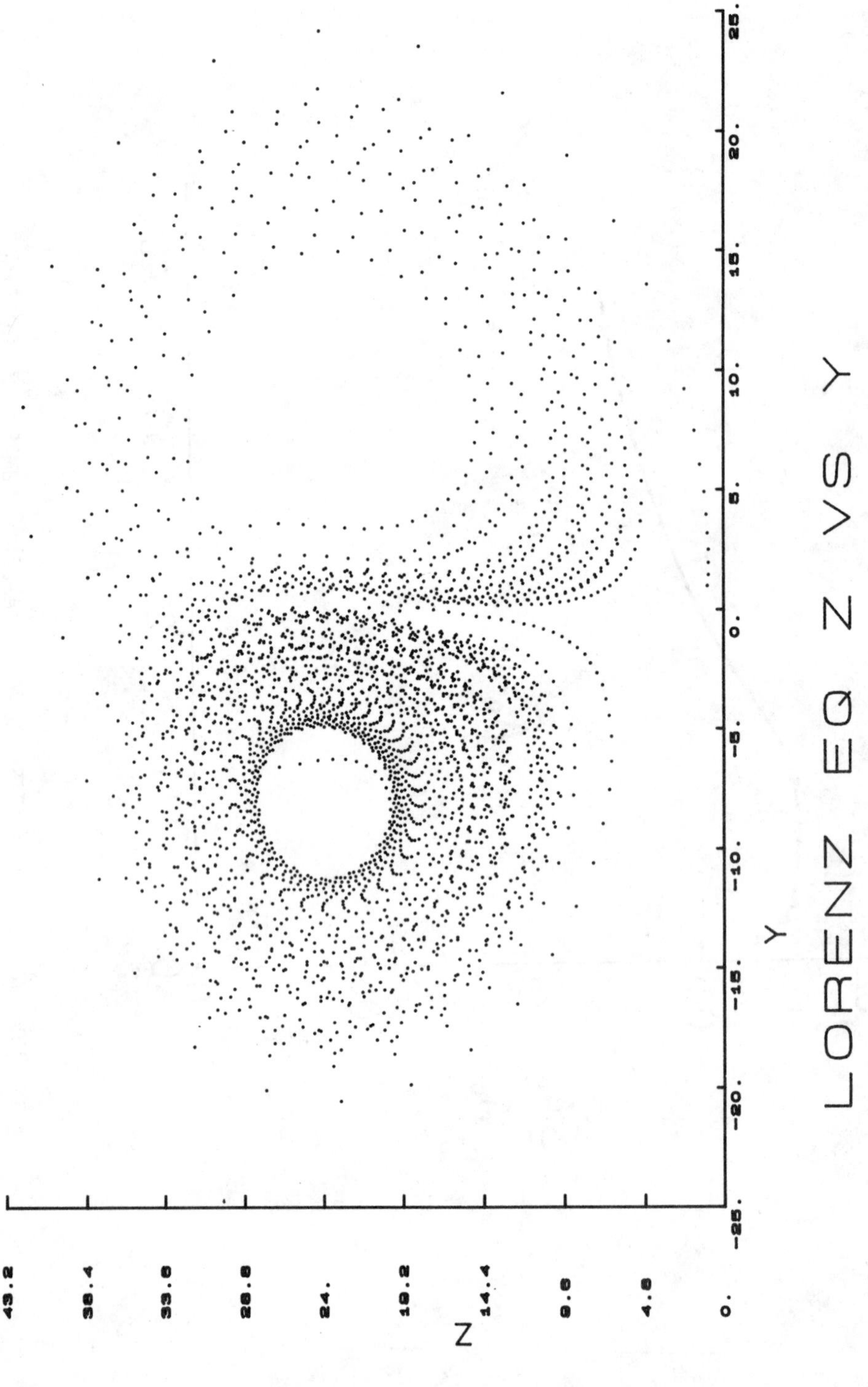

FIGURE 5. Z vs Y; TWO DIMENSIONAL PHASE PLANE TRAJECTORY

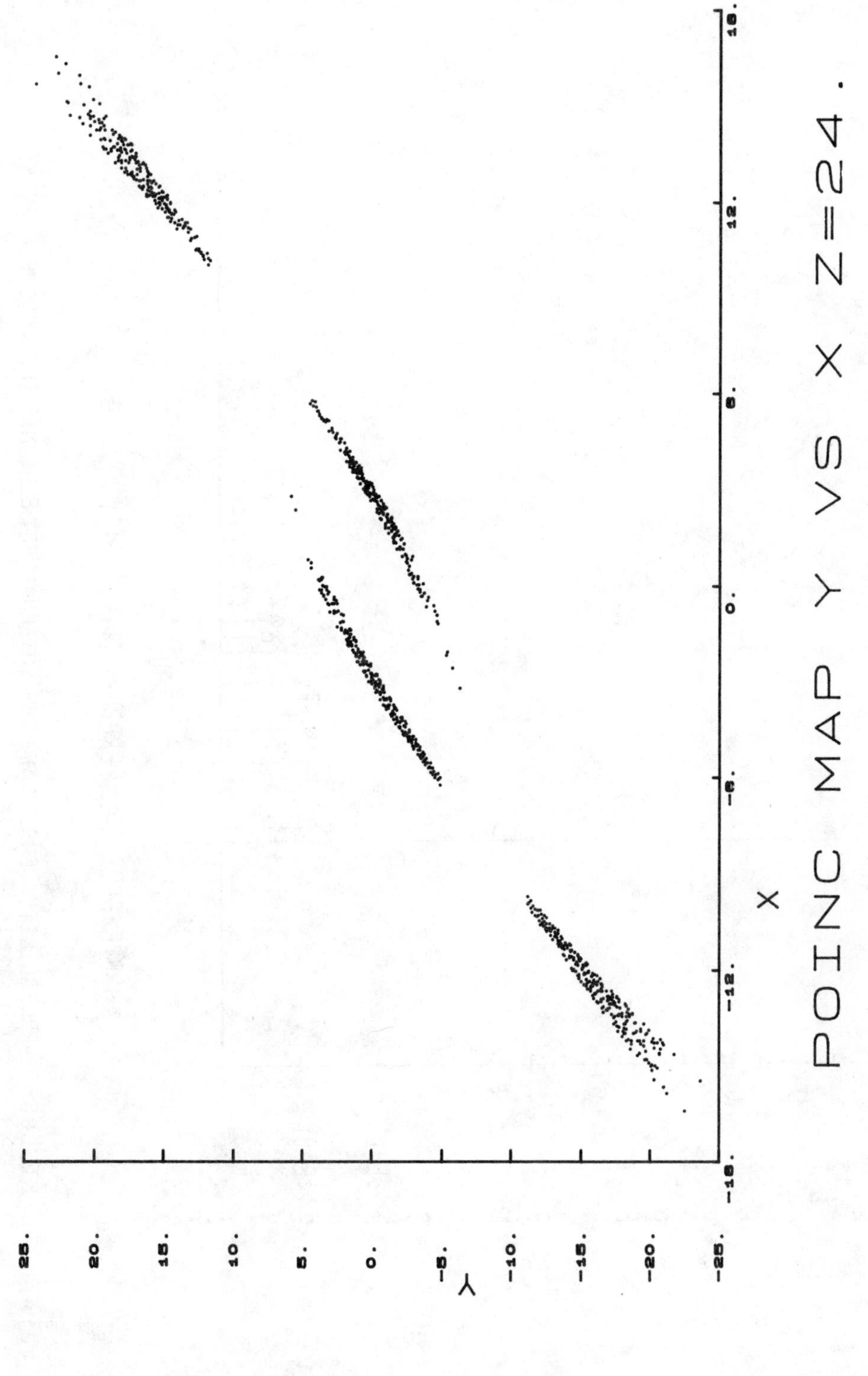

FIGURE 6. POINCARE MAP, Y vs X FOR Z = r-1, r = 25

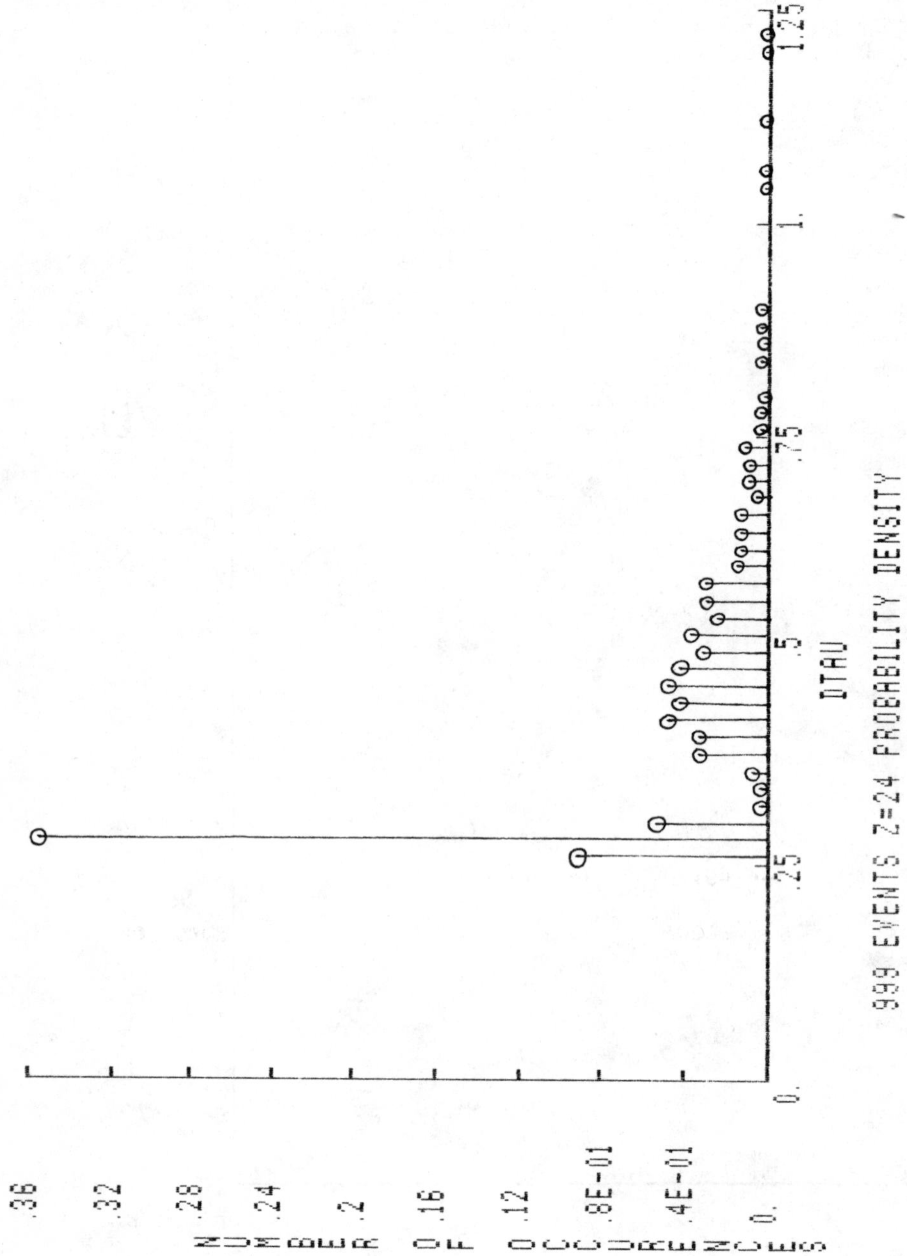

FIGURE 7. PROBABILITY DENSITY OF THE INTERNAL BETWEEN TWO SUCCESSIVE EVENTS; 999 EVENTS, Z = 24

REFERENCES

[1] P. J. Holmes and F. C. Moon, "Strange Attractors and Chaos in Nonlinear Mechanics," Journal of Applied Mechanics, Vol. 108, 1983, pp. 1021-1032.

[2] F. C. Moon, "Experiments on Chaotic Motions of a Forced Nonlinear Oscillator: Strange Attractors," ASME Journal of Applied Mechanics, Vol. 47, 1980, pp. 638-644.

[3] E. N. Lorenz, "Deterministic Non-Periodic Flow," J. Atmospheric Sciences, Vol. 20, 1963, pp. 130-141.

[4] E. H. Dowell, "Flutter of a Buckled Plate as an Example of Chaotic Motion of a Deterministic Autonomous System," J. Sound Vib., Vol. 85, 1982, pp. 333-344.

[5] E. H. Dowell, "Observation and Evolution of Chaos in an Autonomous System," to be published in the Journal of Applied Mechanics.

[6] S. L. Lau, Y. K. Cheung and S. Y. Wu, "A Variable Parameter Incrementation Method for Dynamic Instability of Linear and Nonlinear Elastic Systems," J. Applied Mechanics, Vol. 49, December 1982.

[7] C. Pierre and E. H. Dowell, "A Study of Dynamic Instability of Plates by an Extended Incremental Harmonic Balance Method," submitted to the J. Applied Mechanics.

[8] C. Sparrow, The Lorenz Equations: Bifurcations, Chaos, and Strange Attractors, Applied Mathematical Sciences, Volume 41, Springer-Verlag, New York, 1982.

[9] O. Manley, "Determining Modes and Fractal Attractors of Turbulent Flows," in this volume.